Business Battles in the US Energy Sector

This book is ground breaking in its study of business actors in climate and energy politics. While various studies have demonstrated the influence of business actors across multiple policy domains, this is the first to examine the behaviour of business actors in energy centric industries in the US that will be vital for achieving a clean energy transition, namely the oil, gas, coal, utility, and renewable industries.

Drawing on almost 80 interviews with senior energy executives, lobbyists, and policymakers, it asks two central questions: (i) how and why are business actors shaping energy policy contests in the US? And (ii) what are the implications for policymakers? In answering these questions, this book provides new insights about the preferences and strategies of business in the energy sector, and, significantly, it identifies strategies for policymakers seeking to regulate energy in the face of political resistance from incumbent fossil fuel industries.

This book will be of particular value to students, scholars, and policymakers working in the fields of energy, climate, and environmental politics, as well as individuals generally interested in the role that business exerts over policy processes.

Christian Downie is an Australian Research Council DECRA Fellow in the School of Regulation and Global Governance (RegNet) at The Australian National University, and author of *The Politics of Climate Change Negotiations: Strategies and Variables in Prolonged International Negotiations* (2014).

Routledge Studies in Energy Policy

For further details please visit the series page on the Routledge website:
www.routledge.com/books/series/RSIEP/

Business Battles in the US Energy Sector

Lessons for a Clean Energy Transition

Christian Downie

LONDON AND NEW YORK

from Routledge

First published 2019
by Routledge
2 Park Square, Milton Park, Abingdon, Oxon OX14 4RN

and by Routledge
52 Vanderbilt Avenue, New York, NY 10017

First issued in paperback 2020

Routledge is an imprint of the Taylor & Francis Group, an informa business

British Library Cataloguing-in-Publication Data
A catalogue record for this book is available from the British Library

Library of Congress Cataloging-in-Publication Data
Names: Downie, Christian, author.
Title: Business battles in the US energy sector : lessons for a clean
energy transition / Christian Downie.
Description: Abingdon, Oxon ; New York, NY : Routledge, 2019. |
Series: Routledge studies in energy policy | Includes bibliographical
references and index.
Identifiers: LCCN 2018043514| ISBN 9781138392717 (hardback) |
ISBN 9780429402074 (ebook) | ISBN 9780429687877 (epub) |
ISBN 9780429687860 (mobipocket unencrypted)
Subjects: LCSH: Energy industries–Environmental aspects–United
States. | Energy development–Environmental aspects–United States. |
Energy policy–Environmental aspects–United States.
Classification: LCC HD9502.U52 D69 2019 | DDC 333.790973–dc23
LC record available at https://lccn.loc.gov/2018043514

ISBN 13: 978-0-367-66208-0 (pbk)
ISBN 13: 978-1-138-39271-7 (hbk)

Typeset in Sabon
by Wearset Ltd, Boldon, Tyne and Wear

Contents

Figures

Tables

Preface

It is a fine autumn afternoon in Washington DC in 2014 when I enter a gleaming glass tower to present my photo identification at reception and wait for an escort to the elevator. The sleekness of the building and the absence of metal detectors assure me that the person I have come to interview is not a government employee. But given the story they are about to tell, it does seem that their office would be better located on Capitol Hill.

Sarah[1] is a lawyer, and like many other lawyers in the city she is also a lobbyist. While she has a range of business clients, I have come to talk to her about her work representing energy firms in the oil, gas, and coal industries. Her current assignment is to help business actors in the fossil fuel industries mobilise a coalition to resist President Obama's signature climate and energy initiative the Clean Power Plan. The Plan aims to restrict emissions from power plants under the authority of the Clean Air Act, and many in the business community are not happy about. More importantly, they are mobilising to fight it.

Sarah is charming, generous, and at ease with her assignment, as she explains the types of coalitions that fossil fuel firms are establishing to oppose the Clean Power Plan. She walks me through the regular meetings that are taking place to plot a strategy, the relationships that are being built with other industries, and the resources these actors have at their disposal. This is a sophisticated campaign designed to shape the rules governing billion dollar industries, and she should know, having trod this path before.

As I will show in the following pages, Sarah's story is not uncommon, business actors across the US energy sector are actively engaged in policy contests and there is little doubt that their activities are shaping governance outcomes. This book asks how and why? But why devote a book to analysing business actors in the US energy sector? The answer is twofold. First, if the world is to achieve the clean energy transition that climate scientists conclude is needed then the role of the US will be crucial, given it is the largest producer of oil and gas in the world with the largest reserves of coal on the planet. Second, it is almost impossible to imagine a clean energy transition occurring in the US without overcoming the political resistance of incumbent fossil fuel industries that have delayed, and even derailed, policies designed to encourage clean energy in many parts of the world, including in my home Australia.

In 2015, I moved to the United States to study the behaviour of business actors more closely. I spent countless hours interviewing senior executives and lobbyists like Sarah from all of the major fossil fuel industries, namely oil, gas, and coal. I also had the opportunity to interview executives in the renewable industries who were often battling these incumbent industries. For example, I sat in the offices of major coal corporations and their lobbyists who described the coalitions of business actors across the energy sector lined up to contest a raft of energy regulations designed to limit emissions from coal. I met with executives from oil and gas corporations in Houston and their lobbyists on Capitol Hill, who were campaigning to overturn a 40-year ban on the export of crude oil. I met with small renewable companies that were just starting out, and larger ones with revenues in the billions of dollars that were running endless battles to preserve tax subsidies, which they claimed were only fair given the subsidies to fossil fuels.

By the time I finished researching, the battle lines in the US had been redrawn. President Obama was no longer in the White House; President Trump had replaced him. The impact on climate and energy policy was immediate. Within six months of taking office the new President had announced his decision to withdraw the US from the historic Paris climate agreement signed in 2015. In the months that followed he began reversing course on a series of President Obama's climate and energy policies, which had been designed to restrict the production and consumption of oil, gas, and coal.

In this context it now seems harder than ever to imagine an energy transition occurring in the US. To be sure, many of the people that I interviewed or researched for this book are in a better position to derail attempts at a clean energy transformation today than they were four years ago. Scott Pruitt, until he was forced to resign amid corruption allegations, was perhaps the most egregious example. As Attorney General of Oklahoma, and leader of the Republican Attorneys General Association, Pruitt worked with business actors to fight the Environmental Protection Agency's implementation of the Clean Power Plan. In 2017 he was appointed as the first administrator of the EPA, with the imprimatur of the President and the support of some sections of the business community, to begin the repeal of the Clean Power Plan.

In the current political environment, a clear understanding of how and why business actors shape energy policy in the US is vital. After all, many of the business actors examined in this book could well play a larger role shaping energy policy than they have had under any previous administration. Consequently, the findings of this book are significant not only for those seeking to regulate energy in ways that overcome political resistance from incumbent industries, but also for business actors in the emerging renewable industries that may have to fight to preserve existing policies that encourage clean energy, a topic I return to in the final chapter.

Note

1 Sarah is not her real name. The name has been changed to preserve anonymity.

Acknowledgements

This book would not have been possible without the support of so many people. To begin with, I am indebted to Marc Williams for sponsoring my initial postdoctoral application at the University of New South Wales (UNSW), which provided the funding and space to leave the Federal Government and return to academia to embark on this project. Marc was a constant source of guidance and wisdom, as were many of my colleagues at UNSW. In the same vein so was Peter Drahos at the Australian National University (ANU), whose counsel was critical to shaping the project and narrowing the focus to business actors. I also had the good fortune (Peter may characterise it as misfortune) of travelling to the United States with Peter to conduct the first round of interviews. I am grateful for his encouragement, insight, and humour throughout.

A large part of this project was conceived of in the United States, including in 2015 when I moved to Boston to take up a visiting fellowship at the Massachusetts Institute of Technology. I am sincerely thankful to Lawrence Susskind for once again enabling me to spend time in such an intellectually rich environment while completing the fieldwork. I am indebted to the many people I interviewed across those fieldwork trips that engage with the US energy sector at different levels and from different perspectives. I am grateful for the incalculable hours spent discussing the ideas in this book. Without their insights this project would not have survived.

While in the United States I also benefited from the generosity of countless people who agreed to discuss the project, comment on early chapters, and introduce me to wider scholarly networks. I am grateful to Janelle Knox-Hayes, David Levy, and Stacy VanDeveer, among others, as well as a host of graduate students for their invaluable insights. I have also had the pleasure of presenting parts of this work at conferences and seminars in the United States, United Kingdom, Canada, and Australia. In every case the feedback has helped to shape and improve this book. I would particularly like to thank Robert Falkner who warmly hosted me at the London School of Economics and Political Science in 2016 and whose own work in this field was a source of inspiration.

The final stages of this book have been completed in one of the most vibrant and collegial intellectual homes I have known, the School of Regulation and Global Governance (RegNet) at ANU. I am continually grateful to my colleagues for their warmth and encouragement. In particular, to Neil Gunningham, Kyla Tienhaara, and Susan Sell who provided extremely helpful feedback on different parts of the manuscript, and to Sharon Friel for welcoming me back to RegNet. In addition, Sophie Adams provided superb research assistance, especially in preparing for the fieldwork, as did Alexis Farr, who made managing the final drafting process much easier. This book was improved immeasurably by a number of anonymous reviewers, who took the time to engage with this work. I am grateful to each of them, and to the team at Routledge, especially Hannah Ferguson, for their enthusiasm for this project.

Ultimately, this book, like everything else I do, is only conceivable because of the love and support I am lucky to receive; my partner, Amy, has provided both in abundance, as have my family and friends. I hope by the time our daughter reads this book resistance to tackling climate change will have long been overcome.

Abbreviations

ACC	American Coal Council
ACCCE	American Coalition for Clean Coal Electricity
AEE	Advanced Energy Economy
AFPM	American Fuel and Petrochemical Manufacturers
ALEC	American Legislative Exchange Council
ANGA	American Natural Gas Alliance
API	American Petroleum Institute
AWEA	American Wind Energy Association
BICEP	Business for Innovative Climate and Energy and Policy
BTU	British thermal units
CCS	carbon, capture, and storage
COC	Chamber of Commerce
CRUDE	Coalition (Consumers and Refiners United for Domestic Energy
DoE	Department of Energy
EEI	Edison Electric Institute
EIA	Energy Information Administration
EPA	Environmental Protection Agency
EPSA	Electric Power Supply Association
FTA	free trade agreement
IEA	International Energy Agency
IPAA	Independent Petroleum Association of America
ITC	investment tax credit
LNG	liquid natural gas
LSA	Large-Scale Solar Association
MATS	Mercury and Air Toxic Standards
NAM	National Association of Manufacturers
NEI	Nuclear Energy Institute
NGO	non-government organisation
NMA	National Mining Association
NRDC	Natural Resources Defense Council
OPEC	Organization of the Petroleum Exporting Countries
PACs	Political Action Committees

PACE	Producers for American Crude Oil Exports
PPA	power purchase agreement
PTC	production tax credit
PUC	Public Utility Commission
PV	photovoltaics
SEIA	Solar Energy Industry Association
TASC	The Alliance For Solar Choice
TUSK	Tell Utilities Solar Won't be Killed
US	United States
USCAP	US Climate Action Partnership
WTI	West Texas Intermediate

1 Introduction

Stories from the US energy sector

If the world is to achieve a clean energy transition, the role of the US will be crucial. Not only does the US have enormous global influence, but it is also the largest producer of oil and gas with the largest reserves of coal on the planet. But this is not a book about the US state per se, rather it is about business actors in the US energy sector. This is because without overcoming the political resistance of incumbent fossil fuel industries, it is almost impossible to imagine an energy transition occurring. After all, if policymakers in the US and around the world are to succeed in their attempts to regulate energy and limit greenhouse gas emissions, they will not only need to overcome the resistance of industries that have generated great wealth from burning fossil fuels, but they will also need to build and expand support among renewable energy industries, such as wind and solar power.

In this context, it is important to understand business behaviour. Numerous studies have demonstrated the influence of business actors across multiple policy domains, including in environmental politics. Yet there is less literature on the behaviour of business actors in individual energy-centric industries, namely the oil, gas, coal, utility, and renewable industries (Levy and Kolk, 2002; Meckling, 2011; Skjaerseth and Skodovin, 2003; Newell and Paterson, 1998). And, few studies, if any, have examined the behaviour of business actors in individual energy-centric industries in contemporary policy contests in the US. This book seeks to redress this gap not only to improve our understanding of business behaviour in this critical sector, but also to draw out lessons for policymakers seeking to regulate these industries.

Contemporary policy contests in US energy sector provide an excellent window into business behaviour in the above industries.

* * *

On 7 January 2014, a cold, frigid day in Washington DC, oil and gas executives from around the country gathered for lunch. The occasion was an annual one, the launch of the American Petroleum Institutes' (API) State of the Energy report. Inside the beltway, gatherings like these are a regular affair, but the API is not a regular industry association. It is arguably the most powerful industry association in the most powerful sector of the US economy, the oil and gas industry. Its more than 500 members have combined revenues in the trillions of dollars and include some of the world's largest corporations, such as ExxonMobil, Chevron, and Shell. When the API speaks, Washington listens.

As the attendees sat down to lunch, the speaker was a familiar face, Jack Gerard, President and CEO of the API. Gerard is an old hand in Washington, having led the American Chemistry Council and the National Mining Association prior to his ascendency at the API. This was not his first State of the Energy report, but this year was different. The oil and gas industry was booming. As a result of the shale revolution, the US had overtaken Russia as the largest gas producer in the world and it was now on track to do the same for oil. There was only one problem; the export of US crude oil was banned and had been since 1975 and the OPEC oil crisis (GAO, 2014).

Many in the room were determined to change that. Jack Gerard assured them that the API would lead the charge.

> We will decide if America continues its march toward global energy leadership – a once in a generation choice – or remains content to play a supporting role in the global energy market. We can erase what for decades has been America's greatest economic vulnerability – our dependence on energy sources from other continents, particularly from less stable and friendly nations – and fundamentally alter the geopolitical landscape for decades to come, all while providing a much needed boost to our economy. But only if we get our energy policy right.
>
> (Gerard, 2014)

The right energy policy was to put an end to the de facto ban on crude oil exports. In the months that followed, the API and its members would spend hundreds of millions of dollars making sure that happened. The API was supported by some powerful allies. The next day the US Chamber of Commerce declared its support calling for the ban to be lifted (Mundy, 2014a). Others soon joined, but not everyone was happy. Some in the industry, particularly oil refiners, believed they had much to gain from keeping crude oil on American shores. Exports, they argued, would only increase domestic oil prices and with it their costs of production. But the stage was set; many of the most powerful corporations and associations in the US had begun to mobilise to repeal a law that had been in place for 40 years.

* * *

Mike Duncan, like Jack Gerard, was another journeyman in Washington DC having made his name in the Republican Party in the 1980s and 1990s and ultimately becoming chairman of the Republican National Committee in 2007. Now he was President and CEO of the American Coalition for Clean Coal Electricity (ACCCE), an industry group formed in 2008 to promote coal. But the industry he represented was not in the same shape as oil and gas; coal was not so much booming as busting. In the US coal production is declining, the number of producing mines is declining, productive capacity is declining and the number of employees at coal mines is declining (EIA, 2016a). And thanks in part to the shale boom, its share of electricity generation has fallen from more than half in 1990 to around a third today (EIA, 2017b: 74). The problem for Mike Duncan and the coal industry was that this decline was being accelerated by what they claimed was President Obama's 'war on coal'. On 25 June 2013, fresh from his second inauguration, President Obama launched the latest battle in this so-called war by targeting pollution from coal-fired power plants. Unlike Jack Gerard who was advancing his troops, Mike Duncan was holding the line. A day after the announcement, Mike Duncan went to the offices of the Business Roundtable, which stand in the shadow of the United States Congress on New Jersey Avenue, to deliver his war cry. In attendance were many of the most powerful coal and utility firms in the country, including Peabody Energy and Southern Company.

> Yesterday's news on carbon regulations was a disappointment, but not a surprise. We have seen this threat coming down the road for a while, and yesterday it finally knocked on our door. The President views this as a 'legacy' issue. And on this point, he and I agree. But that 'legacy' is going to be higher energy costs, less reliable electricity, lost jobs and a shattered economy. Even before the President's call for carbon regulations the EPA was extracting pound after pound of flesh from the coal industry.... Our industries can only endure so much. Our economy can only endure so much. The fight before us will come in two stages, one inside the Beltway and one outside. The first round will be fought here in Washington, as public comments are gathered. The second will take place at the state level, as state governments develop plans to meet the proposed standards.
>
> (Duncan, 2013)

It was not the last time Mike Duncan would give this message. Indeed, he would deliver it again, and again, as the war on coal raged and the coal industry fought the administration. Given that coal contributes more greenhouse gas emissions than any other fossil fuel, the battles would have enormous implications for climate change.

* * *

For Rhone Resch it was a battle of a different kind. For more than a decade he had been President and CEO of the Solar Energy Industries Association (SEIA), the peak industry association for solar. Unlike Jack Gerard and Mike Duncan, Rhone Resch was not the same type of Washington journeyman, but he knew the city well enough having served at the EPA in the Clinton administration and worked for the Natural Gas Supply Association. However on 20 October 2014, Rhone Resch was far away from Washington DC, in Las Vegas, Nevada. The reason was the Solar Power International Expo, which his organisation co-hosted. On a warm Las Vegas afternoon, as he stepped out to deliver the keynote address, he was greeted by representatives from hundreds of solar firms, including some of the largest in the world. The solar industry was flourishing in the US. Solar was adding more new capacity than wind. One of the reasons was the Investment Tax Credit (ITC). Established in 2006, it reduced federal income taxes by 30 per cent for capital investments in solar systems on residential and commercial properties. Since the ITC was created, annual solar installations had grown by more than 1,600 per cent, transforming the industry from an $800 million industry in 2006 to a $15 billion one in 2014. Yet the tax credit was due to expire in 2016 and many in the industry feared the worst. The wind industry had been devastated by uncertainty over a similar tax credit the year before. As a result, the solar industry wanted to extend the ITC.

> Today, I'm going to make you a promise: As sure as World War I started in 1914, if the Koch Brothers and their allies come after solar, 2014 will be the beginning of World War III. It's not going to be easy. And, yes, we will be fighting an uphill battle every step of the way.... So today is the official kick-off of our efforts to extend the 30 per cent solar ITC past 2016. Despite the craziness in Washington, D.C., I believe we can win. But being in Vegas should also remind us that we're facing some pretty tough odds again. Make no mistake about it. This absolutely is going to be a long, hard, uphill battle. But by sticking together – and working together – we can be successful once again, just as we were nearly a decade ago.
>
> (Resch, 2014)

Rhone Resch's words may have been exaggerated, but like any good general he was there to rally the troops. While the extension of the ITC may not have been a question of survival for the solar industry, its expiration would no doubt harm it, which is what many in the oil, gas, coal, and utility industries wanted. This was more than just a battle over a federal tax credit. It was also symbolic of a larger war between the new kids on the block in the renewable industries and the incumbents in the fossil fuel industries that had dominated US energy policy for the last 100 years.

* * *

As these examples show, business actors are actively engaged in policy contests across the US energy sector and there is little doubt that they are having an impact. To be sure, numerous studies have demonstrated the influence of business actors across multiple policy domains, yet less work has focussed on the domain of energy (for a review of this literature see Clapp and Meckling, 2013; Tienhaara, 2014). This is somewhat of a surprise given that business actors in the energy sector are central to the problem. In fact, recent evidence shows that just 90 companies are responsible for two-thirds of global greenhouse gas emissions and many of these operate in the US. They include Chevron, ExxonMobil, BP, Shell, ConocoPhillips, and Peabody Energy, all of which have significant US operations. Indeed, together these companies have been responsible for around 13 per cent of all global carbon dioxide and methane emissions since 1751 (Heede, 2014).

Accordingly, in this book, I ask two central questions: (i) how and why are business actors shaping energy policy contests in the US? And (ii) what are the lessons for policymakers? To answer these questions I examine the role of business actors in the oil and gas industries, coal and utility industries, and wind and solar industries across six contemporary policy contests that have taken place during the Obama administration (2009–2016). An exclusive focus on business actors enables a closer analysis of how and why business actors succeeded in exerting influence over the policy process. For example, how and why did oil and gas producers seek to lift the ban on oil exports? How did they seek to frame the contest? Did they lobby, or develop other strategies? And if so, why? Given that resistance from fossil fuels industries to regulation, including oil producers, could delay and even derail government attempts to achieve an energy transition, understanding how these actors behave is of critical importance (Hess, 2014).

Finally, in answering these questions the aim is to build on the empirical insights to identify specific strategies for policymakers seeking to overcome the political resistance of these incumbent industries, and build coalitions in support of policies that encourage the widespread deployment of clean energy, and crudely speaking, reduce the reliance on dirty energy. While the focus is on policy contests in the US, as an energy superpower what happens in the US will have a ripple effect around the world as policymakers in other nations grapple with the same task.

The energy challenge

The global challenge

The climate is changing, and the cause is greenhouse gas emissions. Since the Industrial Revolution, greenhouse gas emissions have increased every year and as a result so too has the temperature of our atmosphere and our oceans. Each of the last three decades has been warmer than any decade since 1850 (IPCC, 2014) and 2016 was the hottest year in recorded history, the third year in a row to record this mark (NASA, 2017). The impacts have been felt around the world including sea level rise, storms, droughts, fires, floods, and famines, not to mention widespread extinctions. Without action, it is projected that global average temperatures will rise by between 4°C and 5°C by the end of the century, rendering parts of the globe uninhabitable (IPCC, 2014).

To avoid the worst impacts of climate change, global temperatures must be kept 'well below' 2°C and ideally below 1.5°C. This is the overarching aim of the Paris Agreement, which was signed in 2015, entered into force less than 12 months later in November 2016, and has now been ratified by 178 nations (UNFCCC, 2018). This is a significant achievement given that it took almost a decade for the Kyoto Protocol to come into force, the last major climate agreement signed in 1997. To achieve this aim, parties to the Paris Agreement have completed national plans – or intended nationally determined contributions (INDCs) – that set out the actions they will take to reduce emissions, such as limiting deforestation or reducing their reliance on coal (UNFCCC, 2015). Many nations, including China, the largest emitter in the world, are on track to meet the targets set out in their INDCs (IEA, 2016b).

However, even if the Paris Agreement is fully implemented, the United Nations estimates that the world will remain on track to increase global average temperatures by 3.2°C by 2100, well above the 2°C limit scientists have warned is necessary to avoid climate catastrophe (UNEP, 2017: xviii). In order to achieve the 2°C target, global greenhouse gas emissions must peak almost immediately and decline sharply to 2100 (Figueres *et al.*, 2017). No easy task, remembering that emissions have risen every year since the Industrial Revolution, and they continue to do so, albeit more slowly. Such is the challenge that most scenarios that seek to limit emissions to below 2°C or 1.5°C assume the deployment of negative emissions technologies. For example, scenarios often combine carbon capture and storage technologies with biomass energy, which permanently remove carbon dioxide from the atmosphere. While such technologies are technically possible, their deployment at scale is untested (Rogelj *et al.*, 2016).

In this context the International Energy Agency (IEA) has long argued that the world needs an 'energy revolution', which results in a rapid transformation to a low carbon system of energy supply (IEA, 2008). As the source of more than two-thirds of global greenhouse gas emissions, it is not hard to see why transforming the energy sector will be crucial (IEA, 2015a). Yet just over 80 per cent of the world's primary energy supply continues to be met by fossil fuels, and, strikingly, this has hardly changed in 40 years. In 2015, oil's share was 31.7 per cent, coal 28.1 per cent and gas 21.6 per cent. Further renewable energy, excluding hydro, has increased from 0.1 per cent of total primary energy supply in 1973 to only 1.5 per cent today – see Figure 1.1.

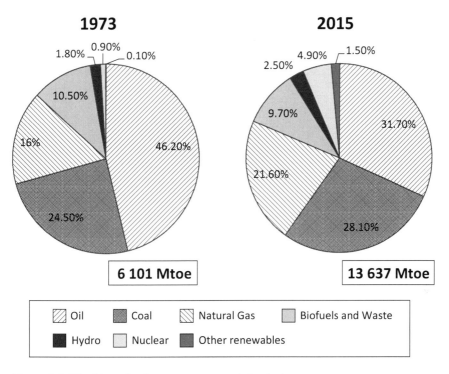

Figure 1.1 World total primary energy supply by fuel.

Source: International Energy Agency (IEA, 2017a).

Recent modelling by the IEA highlights the challenge the energy sector faces to meet the aims of the Paris Agreement. Taking a scenario with a 66 per cent probability of limiting global average temperatures to no more than 2°C, the IEA estimates that the energy sector's carbon budget between 2015 and 2100 – the cumulative amount that can be emitted over that time period – to be 790 gigatonnes (Gt) of carbon dioxide (IEA/IRENA, 2017: 112). As the IEA points out, achieving this would 'require an energy transition of exceptional scope, depth and speed' (IEA/IRENA, 2017: 7). The share of fossil fuels in primary energy demand would halve between 2014 and 2050, while the share of low-carbon sources, including renewables, would more than triple to reach 70 per cent of global energy demand in 2050. As a result, by 2050 almost 95 per cent of electricity would be low-carbon, 70 per cent new cars would be electric, the entire building stock would have been retrofitted and the carbon intensity of the industrial sector would be 80 per cent lower than present (IEA/IRENA, 2017: 8). And all of this is required for just a 66 per cent chance of limiting temperatures to 2°C, when the science shows 1.5°C is required to avoid the most devastating impacts of climate change.

The US challenge

What does this mean for the US? As the second largest greenhouse gas emitter in the world, the role of the US will be central to achieving a clean energy transition. As Figure 1.2 shows, until it was overtaken by China in 2006, the US has been the largest greenhouse gas emitter for many decades, larger than the combined total of emissions from Western Europe. In total, the US contributes around 14 per cent of global greenhouse gas emissions, though its share is declining with the rise of China and India.

While the US position on climate change internationally has waxed and waned in recent decades (Downie, 2014a), under President Obama the US showed an increasing willingness to take action to limit emissions. In 2014 the US, together with China, announced targets for addressing climate change, with President Obama committing the US to reduce its emissions by 26 to 28 per cent below 2005 levels by 2025 (Landler, 2014). In 2016 President Obama, ratified the Paris Agreement adopting the 2014 targets as part of the US nationally determined contribution to the negotiations.

However, recent projections show that even if the initiatives introduced by President Obama were implemented the US would still miss its Paris target by 2025. This includes assuming that the Clean Power Plan is implemented, which was expected to represent half of all emissions reductions contained in current and proposed regulations (Greenblatt and Wei, 2016). With the election of President Trump the emissions

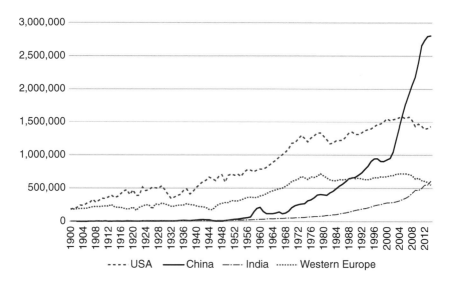

Figure 1.2 Historical CO_2 emissions.

Source: Data from the Carbon Dioxide Information Analysis Center, Oak Ridge National Laboratory (Boden *et al.*, 2017).

challenge will be even harder, especially given that the new President has announced that his administration will not only walk away from the Paris Agreement, but also that it will repeal many of the measures introduced by the previous administration, including the Clean Power Plan (The White House, 2017c).

Irrespective of the ultimate emissions target the US adopts in the future, meeting it will depend on the energy sector (IEA, 2016b: 319). Broadly, the US can be divided into five energy consuming sectors: electricity, industrial, transportation, residential and commercial sectors. Each of these sectors consume primary energy, around 80 per cent of which is supplied by fossil fuels (DoE, 2015). While there are many ways to generate electricity, in the US the sector is largely supplied by coal, natural gas, and nuclear sources, as I discuss below. The industrial, transportation, residential and commercial sectors, also consume most of the electricity generated, though this varies by sector. Of these, the industrial sector is the most diverse. As well as consuming electricity, it also consumes other energy sources directly, especially natural gas and petroleum, to support manufacturing, agriculture, construction, and mining. The residential sector, which includes homes and apartments, and the commercial sector, which includes office blocks, shopping malls, schools, and hospitals, to name a few, also consume electricity as well as natural gas. In contrast, the transportation sector consumes virtually no electricity and is almost entirely dominated by petroleum-based fuels, which are used to fuel cars, trucks, and planes, among others (DoE, 2015). In short, the US energy sector comprises more than simply electricity and there is a range of different energy markets that operate by different rules. For example, business actors in the oil sector operate in a very different environment from business actors in the electricity sector.

Nevertheless, energy-related emissions from these sectors comprise around 80 per cent of total US greenhouse gas emissions (EIA, 2016e: 22). As Figure 1.3 shows, emissions from petroleum have been the largest contributor in recent decades, though they have generally decreased since 2007, despite a recent rise. Coal emissions continue to fall, and have so since the financial crisis in 2007–2009. Emissions from natural gas, on the other hand, have risen since 2009, reflecting its growing share of electricity generation, as falling gas prices have pushed out coal generation (EIA, 2017d: 2). In fact, coal's share of electricity generation has fallen from 54 per cent in 1990 to 34 per cent in 2015. At the same time, non-fossil-fuel electricity generation, which includes nuclear power and renewables, has risen to the point that it equalled that of coal in 2015 (EIA, 2017d: 8). Most of the growth in non-fossil fuel generation has come from wind and solar, which as a share of non-fossil fuels, has grown from less than 1 per cent in 2000 to about 17 per cent in 2015. Whereas electricity generation from nuclear and hydro has fallen over the last two decades (EIA, 2017d: 9).

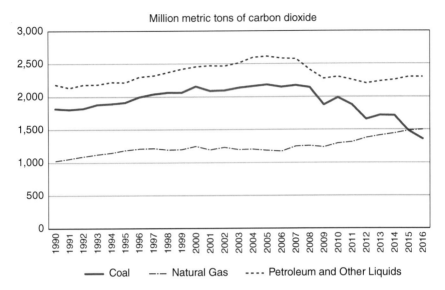

Figure 1.3 US energy-related carbon dioxide emissions by fuel, 1990–2016.
Source: US Energy Information Administration (EIA, 2016d).

Despite the fact that US energy-related carbon dioxide emissions have fallen since 2005 at an average rate of 1.4 per cent annually, they will need to fall much further if the US is to contribute to meeting the 2°C target set out in the Paris Agreement (EIA, 2017b). Recent projections of the US energy sector show that this will not occur without new policy initiatives (EIA, 2017b). The US Energy Information Administration (EIA) projects in its reference case, which assumes that current laws and regulations remain unchanged, such as the Clean Power Plan, that energy-related carbon dioxide emissions will fall by 0.2 per cent annually between now and 2040. In other words, at a slower rate than since 2005.[1] In the electricity sector, for example, it is projected that coal-fired power plants are replaced with new natural gas, solar, and wind capacity, with no significant new nuclear capacity added, as more nuclear capacity is retired than built. Nevertheless this is not enough if the US is to meet its Paris targets. Further, the EIA projects energy-related emissions to be highest in its 'No Clean Power Plan' scenario, which appears the most likely scenario under the current administration (EIA, 2017b).

In summary, if the world is to meet the Paris climate targets, transforming the energy sector will be vital. The energy sector contributes around two-thirds of greenhouse gas emissions. In the US it is starker still with 80 per cent of total emissions energy-related. Accordingly, what is required, as the IEA has argued, is an energy transition of exceptional scope, depth, and speed. Yet the most recent projections show that the world is off track with 80 per cent of primary energy demand continuing to be met by fossil fuels.

The US political environment

Any attempt by the US to take a leading role in achieving an energy transition will not be easy in a political environment characterised by increasing partisanship, especially in the area of climate and energy policy. Since the 1970s congressional politics in the US has shifted away from the norms of cooperation when centrists in both parties regularly cooperated on major issues, to become deeply partisan. This change has coincided with the rise of a dominant conservative faction within the Republican Party, which has squeezed out the moderates, and both parties are now markedly more ideologically homogenous (McAdam, 2017: 195–196). As a result, policy-making in the US is increasingly fraught and many respondents interviewed for this book regularly described a Congress that is 'frozen' or in constant 'gridlock'.

The growing partisanship is especially acute in the area of climate and energy policy where Conservatives and Democrats maintain diametrically opposed visions. In the US, conservatives share a disdain for environmental regulation and view it as a ploy by liberals to impose government control over American life. Conservatives that dominate the Republican Party are deeply sceptical about the ability of the federal government to address social and economic problems and believe that government regulation of the market should be resisted because, among other effects, it reduces individual freedoms, hampers business, and dampens economic growth (Layzer, 2012: Ch. 1). In the area of climate and energy and policy this translates into a resistance against any attempts to address climate change, which has long been viewed as a hoax among conservatives, with renewable forms of power often seen as antithetical to energy security and economic prosperity. In contrast, Democrats generally see a role for government to address the market failure of climate change and promote the transition away from fossil fuels towards cleaner sources of power (Adelman, 2017: 342–343).

The growing dominance of the conservative vision, especially within the Republican Party, is central to understanding the inability of Congress to act in this policy domain. Layzer (2012) argues that although conservatives have not been able to enact wholesale reforms to existing environmental laws, they have been instrumental in blocking efforts to pass major new environmental legislation or increase the stringency of existing laws. They have done so by building a conservative ideological network that since the 1970s has disseminated an anti-regulatory storyline to counter the environmental narrative, mobilised grassroots opposition to environmental regulation and undertaken sophisticated legal challenges to environment laws.

It is no coincidence that at the same time US public opinion has become deeply polarised on the problem of climate change. This was the conclusion of a recent review of polling data in the US, which showed that since the late 1990s opinion on global warming has divided across partisan and ideological lines (Egan and Mullin, 2017). For example, in 1997 equal shares of Democrat and Republican voters said the effects of global warming had already started. Fast forward two decades and the partisan difference on this question and grown by more than 30 percentage points. To make matters worse, action on climate change is not a salient issue, which means there is not a strong constituency for action. Even among Democrat voters that express concern about the issue, climate change is ranked well below many other national priorities. As a result of both the polarisation and low salience, there are often weak incentives for policy-makers to act (Egan and Mullin, 2017).

However, the soaring popularity of President Barack Obama who was inaugurated on 20 January 2009 led many to believe that his presidency could transcend the partisan political divide. In his first State of the Union address, President Obama proclaimed that 'the country that harnesses the power of clean, renewable energy will lead the 21st century' and he promised to increase the supply of renewable energy and work with Congress on 'legislation that places a market-based cap on carbon pollution and drives the production of more renewable energy in America' (The White House, 2009). Yet the reality of the US political environment was reinforced in the following months and years as Republican leaders made a strategic decision to oppose the new President on all issues, and especially climate and energy policy, with the goal of reducing his popularity.

By the 2010 mid-term elections, Republicans had re-taken a majority in the House and significantly reduced the Democratic majority in the Senate. On the back of their success congressional Republicans doubled down on their efforts to oppose the President and they were supported by segments of the fossil fuel sector and conservative advocacy groups, as I will discuss in the following chapters. Little changed in 2012, with Republicans maintaining control of the House and the Democrats control of the Senate. However, at the 2014 congressional elections the Republicans also took control of the Senate for the first time during the Obama administration (US Senate, n.d.). In short, in the aftermath of the 2010 elections the congressional landscape remained extremely difficult for an administration that had professed a desire to advance clean energy and limit greenhouse gas emissions.

Policy contests in the US energy sector

It was in this partisan political environment that business actors battled over key policy contests in the US energy sector. As Table 1.1 shows, in this book the focus is on a series of policy contests across the oil, gas, coal,

Table 1.1 Business battles in key policy contests

The policy contests	General industry positions in contests		Policy outcome
	Support	Opposed	
Gas exports	Oil and gas producers	Petrochemical manufactures	Restrictions on gas exports eased
Oil exports	Oil and gas producers	Oil refiners	Ban on crude oil exports overturned
Emissions trading	Sections of utility industry	Coal producers and sections of utility industry	Emissions trading defeated
Clean Power Plan	Sections of utility industry	Coal producers and sections of utility industry	Clean Power Plan delayed
Production tax credit extension	Wind	Sections of utility industry	Production tax credit extended and then phased out
Investment tax credit extension	Solar	Sections of utility industry	Investment tax credit extended and then phased out

utility, solar, and wind industries. While the outcomes of these contests were directly shaped by the power, preferences, and strategies of business actors, as I will discuss in the following chapters, they were also triggered by the wider transformations taking place in the US energy sector.

First, in the oil and gas industries the so-called shale revolution, which refers to the technology breakthroughs that have enabled producers in the US to access enormous onshore oil and gas reserves, has triggered contests over the exports of these commodities. In the case of gas, a sharp increase in domestic production precipitated a push by gas producers to ease export restrictions, which would enable them to more easily export natural gas to Asia and Europe and thereby take advantage of higher international prices. This sparked a bitter contest across the energy sector, especially from business actors that benefited from the abundance of domestic gas supplies, namely petrochemical manufacturers who did not want it exported. As I will discuss in Chapter 3, by 2016 many of these restrictions had been lifted and the first shipments of US natural gas were on their way to Europe.

The boom in shale gas has been replicated in oil. For decades the US has been dependent on foreign oil, but with a 30 per cent increase in production over the last decade the US has now surpassed Russia and Saudi Arabia as the largest producer of oil in the world (IEA, 2014a).[2] However, for 40 years the US had in place an effective ban on the export of crude oil. Oil produced in the US was to be consumed in the US. With the shale oil boom and surging US production, much like gas, a price spread developed between the domestic price for crude oil – the West Texas Intermediate – and the international price – the Brent. Between 2011 and 2014 the price of the WTI averaged $14 per barrel lower than the Brent (GAO, 2014: 7). Once again, producers pushed Congress to overturn the ban in order to access higher prices for their products on international markets. However, they were resisted by oil refiners that benefited from the price spread. Ultimately, the oil corporations were successful. By the end of 2015, the global oil price had plummeted wiping away the spread, but they had opened the way for future exports, the first of which left for Italy in December 2015 (Carroll and Tobben, 2016).

Coal's enormous contribution to climate change triggered a second set of policy contests in the coal and utility industries. The inauguration of President Obama in January 2009 set off the first contest. With the President's support Congress attempted to legislate a nationwide emissions trading scheme, which would set a cap on greenhouse gas emissions especially those from the electricity sector. The focus on the electricity sector reflected the fact that more than half of the coal produced in the US is used to provide electricity, and this increases to 90 per cent for steam coal (Witter, 2015c: 4). As a result, business actors in the coal industry, in particular, mobilised to oppose the legislation, while the utility industry split in part reflecting the varying use of coal in their generation portfolios.

The failure of this legislation led to a second attempt three years later, which was just as contested and led to similar divisions within and between industries. This time around President Obama directed the EPA to establish carbon pollution standards, with the aim to reduce emissions from power plants by 30 per cent from 2005 levels by 2030 (The White House, 2013b). Following the release of the proposed rules in 2014, the plan was finalised in August 2015 (EPA, 2015). However, the US Supreme Court granted a stay in February 2016, stopping the implementation of the Plan, which is now unlikely to be implemented in its current form with the election of President Trump (Adler, 2016).

A third set of policy contests surrounds the fact that in the US and around the world renewable energy is booming. This has set off a battle between the traditional incumbents in the energy sector and their rivals in the renewable energy sector, particularly in the wind and solar industries, over federal tax credits. In the case of the wind, one of the most important tax credits has been the production tax credit (PTC), which for wind was $23 per MWh in 2015 (DoE, n.d.) The PTC has been vital for the development of the wind industry, despite the fact that it has been renewed and revised multiple times creating significant uncertainty. Fortunately for the wind industry in 2015 they were able to secure an agreement that extends the PTC to 2019.

This contest was closely tied to a parallel contest in the solar industry over the ITC, which reduces federal income taxes by 30 per cent for capital investments in solar systems on residential and commercial properties (SEIA, 2015c). The ITC has been a boon for the industry since it was established in 2006, with dramatic increases in utility scale and small scale solar. As part of the same agreement that extended the PTC in 2015, the solar industry secured the extension of the ITC for another six years to 2022. As I will discuss in Chapter 5, these contests appear to be part of a larger war between incumbents in the fossil fuel industries and the new kids on the block in the wind and solar industries.

Business actors

Although the next chapter will draw on the existing literature to provide a theoretical basis for examining business behaviour, it is useful to say a little more here about business actors in the social sciences. First, what are business actors?[3] In the social science literature on non-state actors a general distinction is made between for-profit actors and non-profit actors. In essence, this is a distinction based on motivations. Business actors are for-profit actors and are primarily motivated by instrumental goals, normally the pursuit of profit for their owners or shareholders. Non-profit actors, on the other hand, such as environmental NGOs, are not and typically lay claim to a common good. It goes without saying that such distinctions are never perfect, and some scholars have challenged this distinction based on instrumental motivations (Sell and Prakash, 2004). In this book the focus is on for-profit business actors, rather than on the role played by non-profit actors or by state actors, though as I will discuss in Chapter 2, to the extent that these other actors interact with business actors they are considered in the analysis. In short, the focus is on actors in the oil, gas, coal utility, solar, and wind industries

Over the last half a century scholars across the social sciences have shifted from an exclusive focus on the state to examine the role of non-state actors, including business actors. Traditionally in political science business actors have been viewed in the pluralist tradition, where corporations compete for influence just like any other interest group at the domestic or international level. This perspective has long dominated studies of US politics where scholars argued that corporate actors do not possess any advantages that are not held by other interest groups (Dahl, 1961).

However, in recent decades scholars have challenged this conventional understanding. Political scientists have turned their attention to the power of business actors and their capacity to shape policy outcomes (Schattschneider, 1960; Vogel, 1989; Culpepper, 2015). Business and management scholars have examined the political activity of firms, their non-market strategies, and the impact these have on firm performance and industry competition (Shaffer, 1995; Håkansson and Ford, 2002). Parallel work in regulation and governance, and in the tradition of public policy, has confronted the enduring consensus that governance outcomes are the product of state actors operating through formal hierarchies, to show that business actors often govern in concert with networks of state and non-state actors across multiple levels of governance and multiple policy domains (Braithwaite and Drahos, 2000; Rhodes, 2006; Sabatier, 1988).

As I will discuss in the next chapter, existing research in these fields shows that business has played a critical role shaping governance outcomes around the globe. In global environmental politics scholars have considered the complex means via which business actors have shaped environmental policy at the national, international and transnational level (see for example, Falkner, 2008; Clapp and Fuchs, 2009; Levy and Newell, 2002). Other scholars have explored these issues, for example, in relation to intellectual property (Sell, 2003; Sell and Prakash, 2004). Significantly, this literature has not yet examined the energy sector in any depth.

Cases, methods, and data

This book examines the role of business actors in the energy sector across six policy contests during the Obama administration (2009–2016). The focus is on the US because if the world is to achieve an energy transition the US will be critical given their historical influence in writing the rules that govern the globe (Braithwaite and Drahos, 2000). And, within the US the focus is on business actors in incumbent fossil fuel industries whose political resistance must be overcome if an energy transition is to be achieved, and it is on business actors in renewable energy industries whose political support for clean energy will need to grow. Accordingly, industries engaged in the production and consumption of oil and gas are examined because, as discussed, the evidence shows that a third of the world's oil reserves and half the world's gas reserves must be left in the ground to keep global warming to 2°C. Industries engaged in the production and consumption of coal are considered for the same reason because almost 90 per cent of global coal reserves must be left untouched. And, industries that produce and consume wind and solar power are examined because most projections show that a rapid widespread deployment of these renewable technologies will be necessary to replace these fossil fuels.

In order to examine business actors, the largest firms were identified in each industry according to publicly available data. For the purposes of this study the industry is unit of analysis. In the oil and gas industries producers were identified based on annual revenues sourced from the Global Fortune 500 lists, where this was not available data was sourced from company annual reports or associated industry reports.[4] In the coal industry the largest coal producers were identified based on production data sourced from the EIA and in the utility industry data were sourced from the Edison Electric Institute (EEI), the industry association for investor-owned utilities (EIA, 2014; EEI, 2014). Finally, in the wind industry manufacturing firms were identified based on their share of the US wind power market, and in the solar industry firms were identified based on their share of the manufacturing and installation segment of the US solar market (Osten, 2015; Khedr, 2015; DoE, 2016). To be selected a firm did not need to be headquartered in the US, but it did need to have a presence in the US market. For example, Royal Dutch Shell has its headquarters in the Netherlands, but it remains an active participant in US policy debates. The same approach was taken in each of the industries.

While there is a growing trend for large-N quantitative studies in the social sciences (Gerring, 2017), they tend to lean toward examining factors that are measurable and neglect those that are more difficult to quantify, such as mapping networks of diverse actors (O'Neill *et al.*, 2013: 449). This book primarily employs a small-*n* case study approach. This was considered the most effective method for analysing the how and why of business behaviour. I used process tracing to reconstruct chronologically the behaviour of business actors in these industries, particularly their strategies, in each of the six policy contests (Bennett, 2007: 35–36; Beach and Pedersen, 2013). The six policy contests were chosen to vary across the key energy industries that the empirical evidence indicates are likely to determine the speed and scope of an energy transition in the US, as noted they are the oil, gas, coal, utilities, wind, and solar industries. As a result, it was possible to examine how preferences and strategies varied across these key industries.

At the same time, the policy contests chosen shared some fundamental characteristics that allowed a number of potential variables to be held constant, which are theoretically important for analysing the behaviour of business actors (Przeworski and Teune, 1970: 32–34). First, the policy contests all took place in the US. As a result, business actors across the policy contests all faced the same opportunities and constraints from operating in the same political system. For example, they all had the same opportunity to form coalitions with the same sets of state and non-state actors, to lobby and shape regulations. The conditions would be very different, for instance, for energy corporations operating in China or Russia. Second, each of the policy contests took place at the Federal level. Hence business actors were interacting with similar sets of policymakers in each of the contests, namely the White House, the Congress, and Federal agencies, such as the Department of Energy (DoE). In other words, policy contests at the state level or local level, such as those over renewable portfolio standards were excluded. That is not to say that business actors never looked to other venues to make their case, they did, but they did so only to the extent that it could help to shape a federal policy outcome (Klyza and Sousa, 2008). Third, the six cases all occurred during the period of the Obama administration (2009–2016). Previous research has shown that a change in administration can significantly affect the capacity of non-state actors to shape governance outcomes (Downie, 2014b). As a result, cases were selected to ensure that this variable remained constant, although there were changes in the composition of the Congress during this period, as discussed above.

Further, cases were also selected that respondents and empirical evidence indicate are significant in their own right because of the impact they are having on the US energy sector. For example, respondents in the coal industry argued that the contest over the Clean Power Plan is having significant impact on the viability of coal, likewise respondents in the solar industry argued the same about the PTC, and the evidence supports this. Finally, each of these cases were 'policy contests'. In other words, as respondents indicated, business actors in the energy sector were actively engaged in these contests during the Obama administration. This has two benefits. On the one hand, it makes it possible to identify and examine business preferences and strategies relative to other policy debates where business actors were less active. On the other, because these policy contests have a high level of visibility it also makes it easier to ascertain the domestic political incentives of policymakers at a general level given the publicly available information.

In order to examine business behaviour in each of the six policy contests semi-structured interviews were conducted with business actors. Specifically, three rounds of semi-structured interviews were conducted with senior executives and lobbyists from energy corporations and industry associations, supplemented with a small sample of policymakers and academic experts. The first round of interviews was conducted in 2014 and was used to help identify the participants in the contests. In particular, to identify business actors that were not captured in the original sample, but that respondents claimed were important players in the policy contests despite, in some cases, their smaller size. Subsequently, two further rounds of interviews were conducted concentrating on firms engaged in these policy contests. In order to access these firms more easily I was based at the Massachusetts Institute of Technology in Cambridge, MA, in 2015. This provided not only an excellent vantage point from which to examine US energy policy, but it also enabled multiple trips to the headquarters of firms in these industries, including their political headquarters in Washington DC. In all, 76 respondents were interviewed. While interviews were conducted confidentially, most respondents agreed to be cited as representatives of their industry rather than of a specific firm or association.

To ensure construct validity, the data from the interviews were analysed in two ways. First, the data were evaluated for consistency within each policy contest. Data provided from one firm in one case were checked against the data provided by his or her colleagues in the same industry. This is especially important in elite interviewing because of the risk that respondents may exaggerate the importance of their role in events (Berry, 2002; Delaney, 2007). Second, data from the interviews were compared against documentation collected from an earlier literature review in which a firm's position or strategy in each of the policy contests were revealed. Such documentation included congressional testimony, press releases, speeches, or newspaper reports covering the firm's behaviour, or similar documentation from industry associations, which outlined the industry's position and approach (Cass, 2007; Bennett, 2007). Once this validation process was complete a rich body of empirical data was available for analysis.

Contributions of the book

This book begins from the premise that the world needs a clean energy transition and that the US is crucial to making that happen. From this starting point I argue that it is hard to imagine an energy transition occurring in the US without overcoming the political resistance of incumbent fossil fuel industries. A clear understanding of business behaviour in the energy sector therefore is a necessary first step to achieving a clean energy future. In the absence of knowing exactly how and why business actors behave in the US, government attempts to transform the energy sector are likely to be delayed, or even derailed, by the resistance of actors in the incumbent fossil fuel industries.

In taking this first step, this book makes an empirical, theoretical, and policy contribution. Before canvassing what those contributions are, it is helpful to be clear about the scope of the argument. In other words, to be clear about what this book does not do. This book does not seek to explain policy outcomes vis-à-vis other actors, which would necessarily require consideration of a wider range of state and non-state actors. In other words, I do not claim to provide a definitive explanation for the outcome in each of the policy contests. Instead the focus is exclusively on business actors in key energy industries. That said, given that existing empirical studies have shown that business actors influence the ultimate shape of policy outcomes, I do attempt to assess the impact business actors had on each contest through an analysis of their preferences and strategies. Of course, business behaviour does not occur in a vacuum, it is shaped by the institutional environment in which it operates. Accordingly, as I explain in the next chapter, I consider two conditioning factors that are likely to be critical to the policy contests examined in this book: the mobilisation of other non-state actors and the role of policymakers.

The principal empirical contribution of this book is that it examines business actors in the energy sector, a topic that until now has been largely neglected in energy politics and environmental politics. Existing scholarship has shown that business actors are critical to addressing some of the most pressing environmental problems facing the globe, but little if any scholarship has considered the role of business actors in energy-centric industries, despite the fact that energy contributes around two-thirds of global greenhouse gas emissions. Accordingly, this book is the first to provide a fine-grained empirical analysis of these actors in contemporary policy contests in the US. As a result, it maps the key actors, coalitions, and networks, and identifies not only their preferences, but also the strategies they use to shape outcomes. In addition, because the political histories of these contemporary policy contests are yet to be examined in detail in the existing literature, the accounts provided in the following chapters make an important empirical contribution.

Second, in doing so, this analysis provides new insights about the preferences and strategies of business actors in the energy sector. First, it speaks to existing theoretical expectations about the factors that drive firm preferences, including commercial interests, and to the factors that shape how firms respond to the institutional contexts in which they are embedded. It also speaks to expectations that variations in the distributive effects of policies on business actors will lead to divergent preferences and often industry conflict. While it confirms much of the existing theory, it also suggests new insights about how business preferences are determined, especially about how business actors hedge their positions, which appears to be especially prevalent in some industries. Second, this book builds on the existing literature to offer new theoretical insights about the strategies business actors use to exert influence over the policy process. For example, in the context of coalition building, a key business strategy, the empirical evidence highlights how traditional industry associations often act as the command centre of business campaigns to pool resources and build legitimacy. It also highlights how ad hoc coalitions emerge and are prevalent across the sector, which is often overlooked in existing studies, and it highlights the important role that coalitions can provide in building the legitimacy of emerging industries, namely renewable industries. In addition, it considers the influence business actors had on each of the policy contests, and importantly, how their opportunities to do so were affected by the role of other non-state actors and by policymakers, whose own behaviour was driven by their beliefs and political incentives.

Third, in the final chapter this book seeks to make a policy contribution by drawing together the theory and the evidence to identify specific strategies for policymakers seeking to facilitate a clean energy transition. Strategies are identified in the context of the US energy sector for policymakers concerned about implementing polices that can overcome the resistance from incumbent fossil fuel industries, and that build coalitions and networks in support of policies that promote clean energy. In doing so, it builds on recent work that explores the pathways to building winning green coalitions. Specifically, the strategies are to entrench and build existing interests via targeted sector specific policies; exploit inter-industry and intra-industry divisions with smart policies that, for example, target politically weak industries; and shift existing interests with policies that induce changes in industry investment and structure by sending direct and repeated policy signals. Further, following Donald Trump's election, strategies are also identified for business actors in emerging renewable industries, which could be employed in the absence of attempts by federal policymakers to advance an energy transition.

Guide to the book

This book is organised into three parts. The first part – Chapters 1 and 2 – introduces the transformations taking place in the US energy sector and the role of business actors therein. Chapter 2 provides a theoretical basis for examining business actors in key industries in the six policy contests outlined in this chapter. Drawing on the insights of scholars in political science, business and management, and regulation and governance, it establishes a theoretical framework for considering the power, preferences, and strategies of business actors. While the focus is on business actors, this chapter also considers conditioning factors that will impact the influence business actors can exercise, namely the mobilisation of other non-state actors and the role of policymakers.

The second part of this book is organised around critical energy sources: oil and gas (liquid fuels); coal and utilities; and wind and solar power. Each chapter – Chapters 3, 4, and 5 – considers the role of business actors in two policy contests, with a particular focus on the preferences and strategies of actors in key energy industries. Chapter 3 examines business actors in the oil and gas industries using two policy contests triggered by the shale revolution, which has resulted in the US becoming in a matter of years the largest producer of oil and gas in the world. That is, the contest over the export of gas and the contest over the export of oil. Both provide an excellent window via which to examine business in these incumbent fossil fuel industries that have long dominated the US energy sector.

Chapter 4 turns to coal to examine business actors in the coal industry that mine the coal and in the utility industry that burn it to generate electricity. While the US continues to have the largest estimated recoverable coal reserves on the planet, demand for coal has been declining largely because the electricity sector, which consumes almost 90 per cent of US coal production, has been substituting coal for cheap gas (EIA, 2016f). In this context, President Obama's attempts to implement an emissions trading scheme, and later the Clean Power Plan, two initiatives that are directly designed to limit the exploitation of coal, have been fiercely contested by business actors in both the coal and utility industries.

Chapter 5 continues the examination of business actors, though in this chapter the focus is on business actors in the renewable energy sector. Of all the renewable sources that are commercially viable today, wind and solar have the greatest potential to transform the energy sector. In the US wind and solar power have surged in recent years precipitating a series of policy contests, particularly with the utility industry, which is directly threatened by the spread of renewable power. Two of the most hotly contested policy contests have been over the extension of the PTC and the ITC. In both cases business actors in the wind industry and the solar industry have been directly engaged, as have other business across the US energy sector.

Following an analysis of business actors in each of the policy contests, the third part of the book – Chapters 6 and 7 – draws together this empirical work to synthesise the theory and evidence and identify lessons for policymakers seeking to regulate these industries. Based on the empirical analysis, Chapter 6 distils the theoretical insights about why business actors behave the way they do, that is their preferences, and how they behave, that is, the mechanisms via which they seek to exert influence over the policy process. In doing so, it shows the importance of understanding the commercial interests of actors and their institutional environment. It also shows how energy corporations build coalitions, frame debates and coordinate lobbying activities to advance common goals. And finally, it reflects on the influence of business actors and how their opportunities to shape outcomes have been affected by the mobilisation of other non-state actors and the role of policymakers.

Chapter 7 asks the question: what should policymakers do? It identifies strategies that policymakers can employ to implement policies that can overcome the resistance from incumbent fossil fuel industries. It also offers lessons for business actors in renewable industries faced with a new political reality following the election of Donald Trump to the White House.

This chapter concludes by reflecting on the implications for future research and the implications for the climate of the energy transformations taking place in the US. It highlights that while the structural decline of the coal industry and the boom in the wind and solar industries have obvious benefits for efforts to limit global greenhouse gas emissions, the US much like the rest of the world, remains off track when it comes to the oil and gas industry.

While I have made every attempt to keep this book as slim as possible, for those short of time this book can be read in parts. Readers with a specific interests in the oil and gas sector (Chapter 3), coal and utility sector (Chapter 4) or wind and solar sector (Chapter 5), may wish to select only the relevant chapter, though I suggest those with an interest in electric utilities read both Chapters 4 and 5. For policymakers that have little interest in the theory I recommend jumping straight to Chapter 7. That said, I hope that scholars engaged in business actors and energy and environmental politics will find enough interesting material to read from the first page to the last.

Notes

1 These projections are based on current laws and regulations implemented during the Obama administration, including the Clean Power Plan, and do not take into account recent announcements by President Trump to roll back such measures.
2 US oil production figures include crude oil and natural gas liquids. Based on crude oil alone the US remains the third largest producer in the world.
3 I will use the terms business actors and corporate actors interchangeably throughout the book.
4 Details of the Fortune 500 methodology can be found here: http://fortune.com/fortune500/

2 Understanding business preferences, strategies, and influence

Introduction

Business actors have been central to American politics. In the domain of energy, firms, such as ExxonMobil and Chevron, that trace their history back to Standard Oil, have played critical roles setting political agendas, influencing policymakers, and shaping policy outcomes (Sampson, 1975). Since the 1950s and 1960s social scientists have turned their attention away from an exclusive focus on the state, to consider the behaviour of business actors. Writing in 1960, Eric Schattschneider (1960: 116) argued that 'once upon a time the church was the principal nongovernmental institution; today it is business'. In the decades since, social scientists have produced a substantial body of literature on the role of business actors in political processes. Political scientists, for example, continue to debate the sources of business power and whether it occupies a privileged place in society (Culpepper, 2015). Business and management scholars have turned their attention to business political activity, or nonmarket strategies, and the impact these have, for example, on firm performance and industry competition (Shaffer, 1995; Hillman *et al.*, 2004; Mellahi *et al.*, 2016). Regulation and governance scholars, and those in the tradition of public policy, have overturned the long held contention that governance outcomes are the product of state actors operating through formal hierarchies. Significant empirical work now shows that policy processes are influenced by networks of state and non-state actors operating at multiple levels of governance, across almost every policy domain imaginable from finance, to trade, and the environment (Braithwaite and Drahos, 2000; Sabatier, 1988).

However, there remains much we do not know about the role of business actors. As David Vogel (1996) argued in the 1990s, and it remains true today, the study of business actors has not emerged as a central focus of study in these disciplines. For example, very few political scientists teach at business schools, and business has not developed into a distinctive subfield of political science. The lack of attention to business actors is especially evident in the field of energy. As discussed in Chapter 1, while scholars of environmental politics have built upon the theoretical and empirical insights of these literatures to study business actors, especially in area of climate change, key questions remain about business actors in energy centric industries in the US. What role are business actors playing in the US energy sector? How and why are they seeking to shape policy contests over energy? In particular, what are their preferences? What strategies have they employed? Answering these questions will also help to answer subsequent questions about the influence of these actors on US energy policy and about the lessons that can deduced for policymakers seeking to regulate the energy sector.

In this chapter I draw on these literatures, and synthesise their insights, to provide a theoretical framework for examining business actors. As discussed in Chapter 1, the focus of the analysis is on key energy industries with the largest firms in each industry identified according to publicly available data. This contrasts with much of the writing in political science, though, of course, not all, including environmental politics, which treats 'business' as a single entity and not the various firms, industries, and associations it comprises, and it contrasts with the business and management literature, which is distinguished by its use of the firm as the dominant level of analysis (Shaffer, 1995: 497; Prakash, 2000). In setting out the theoretical framework, the aim is to provide an introduction to the theoretical literature on business actors in the social sciences, and, importantly, to consider the key variables that are expected to determine the preferences of business actors, which in turn inform the strategies they use to shape outcomes.

The next section provides an introduction to the power of business actors in US politics. This is followed by a discussion of business preferences and then business strategies. The penultimate section considers conditioning factors that are likely to create and limit the opportunities for business actors to influence policy contest. The chapter concludes with a summary of the theoretical framework that will be used to guide the empirical research in the subsequent chapters.

The power of business actors

The power of business actors in American politics has been an ongoing point of contention. After World War II, political scientists, largely in the pluralist tradition, tended to view business actors as just one of many interest groups competing for influence. Pluralists argued that business actors do not possess any advantages that are not held by other interest groups (Dahl, 1961). Within this tradition, scholars claimed that business therefore possessed far less power to dominate political processes than was popularly associated with them (Truman, 1953).

However, in the 1970s debates about business power exploded in the social sciences and many scholars critiqued the pluralist position arguing that business enjoys a privileged position in American politics. The most influential critique arguably came from Charles Lindblom, who repudiated his earlier position as a pluralist, with the publication of *Politics and Markets* in (1977). In essence, these scholars argued that business power was not just a function of its superior economic and political resources, but instead it enjoyed a privileged position in the capitalist system because of its contribution to the economy. In other words, business possessed a type of structural power, which was unique to business actors. Similar debates raged among Marxist scholars, with prominent Marxists of the 1970s, such as Ralph Miliband arguing that political power was concentrated in the hands of the dominant business class (Miliband, 1969). And among economists too, for example, in the public choice literature scholars debated the role that interests groups can play in capturing government agencies to advance their own interests, so-called regulatory capture (Stigler, 1975).

The pluralist position may have been discredited by the late 1970s, but the structuralist critique had its own problems and fell out of favour in the following decades. In the 1980s David Vogel argued that instead of being relatively stable, business power tends to fluctuate with time and the political preferences of those in power. Hence whether business is powerful depends on the period one is interested in (Vogel, 1989). Since the 1980s to today there is good reason to assume that the period has been one that has tended to favours business. For example, Blyth and Matthijs (2017) argue that since the 1980s the institutions upon which policies are developed and implemented, what they refer to as the 'macroeconomic regime', have shifted from focussing on achieving full employment prior to the 1980s to achieving price stability since. As a result, some actors, largely business actors, are empowered by the target of price stability, which is to be achieved via flexible markets and limited government intervention, and some actors are disempowered, largely workers and unions.

Similar debates have played out across different disciplines for several decades. Scholars have variously claimed that the power of business is in ascendency or decline. For instance in the 1960s and 1970s, many scholars argued that corporations were constraining state power, especially in developing countries, and that the sovereignty of states was 'at bay' (Vernon, 1971). In the 1980s, the growth of economic nationalism led others to suggest a resurgence of the state and to raise questions about the sovereignty at bay thesis (Biersteker, 1980). In the decades since, these ongoing debates about the rise and fall in the power of large corporations with respect to the state have continued. The transformations wrought by globalisation in the 1990s once again led many scholars to describe a borderless world in which global corporations are the primary actors. Within the tradition of international political economy there was a widespread view that the state was in 'retreat' (Strange, 1996), yet others have argued persuasively for 'bringing' back the state (Weiss, 2003; Morgan and Orloff, 2017).

These binary debates are not always helpful because they assume that business power and state power are a zero sum game. This is clearly not the case. Nor does it tell us anything about the types of power. Instead, it is more helpful to take the view that Barnet and Muller (1974: 74) did in one of the earlier works on large corporations when they argued that 'the rise of the modern corporation has been paralleled by an extraordinary rise in the activities of the modern nation state'. This view of the state and business evolving together and exercising power in relation to each other rather than at the expense of each other, leads to more fruitful questions about power, such as what is power? What powers do business actors possess? And what are the dimensions of that power? These questions are important because the power of business actors in key energy industries can affect the choice of strategy and the effectiveness of a chosen strategy, as I will discuss below.

Building on the literatures that have followed the pluralist debates in recent decades, Barnett and Duvall (2005: 3) argue 'power is the production, in and through social relations, of effects that shape the capacities of actors to determine their own circumstances and fate'. This definition is useful because it highlights the multiple dimensions of power beyond what is popularly conceived. Broadly speaking, scholars in political science, including scholars in environmental politics, who study business actors, have identified three principal dimensions of business power: instrumental, structural, and discursive power, although Barnett and Duvall (2005) discuss four.

First, instrumental power (sometimes referred to as relational or compulsory power) refers to the direct power that one actor has over another to shape directly the circumstances or actions of that actor (Barnett and Duvall, 2005: 13; Clapp and Fuchs, 2009: 8). This type of power is often described with reference to Robert Dahl's (1961) definition of power, that is, the ability of A to get B to do what otherwise B would not do. The ability of A reflects A's material resources. In the context of business actors, it typically reflects their financial resources, which they can use to engage in activities to influence policy outcomes, such as lobbying, which will be discussed below. Similarly, in business and management, scholars focus on the resources and capabilities of firms to control their political and economic environment to improve their economic performance. Thus they consider the capacity of business actors to assemble and leverage resources, including management expertise and financial commitments, to pursue nonmarket strategies (Mellahi *et al.*, 2016; Doh *et al.*, 2012).

In the 1970s the challenge to the pluralist perspective and the definition of power as instrumental power led scholars to propose the concept of structural power, a second dimension of business power. Structural power emphasises the structural positions in society and the power it confers (Bowman, 1996: 16). According to Barnett and Duvall (2005: 18–19), structural power shapes the fates and conditions of existence of actors in two ways. First, structural positions generate unequal social privileges and capacities to actors, such as capital to labour. Second, drawing on Luke's (1974) work, they argue that it can also constrain some actors from recognising their own domination. It is these ideas that underpin Gramscian and historical materialist theories in political science and international relations (Gill and Law, 1989; Levy and Newell, 2002).

However, as Culpepper (2015: 392) points out in a special issue in *Business and Politics*, structural power remains a 'clunky variable'. Part of the problem is observing it in practice and distinguishing it from instrumental power. For example, if coal producers and electric utilities succeed in scuttling environmental regulations, does this reflect their structural power in the economy or their instrumental power to lobby policymakers? The question then follows, is it even a helpful distinction? In the same special issue, Young (2015) finds that firms seem to do best when they are both 'structurally prominent' in his words, and when they are also actively engaged in the exercise of instrumental power. In other words, it may be more helpful to see how both dimensions of power work together.

While political scientists have typically focussed on instrumental and structural power, a third type of power, which draws on sociological traditions in the literature can also be identified: discursive power. Unlike instrumental power, and especially structural power, which are generally associated with the position of business actors in society, discursive power is associated with the ideas and values that infuse the exercise of power (Schmidt, 2010). Discourses are powerful therefore, because they can be used by actors to frame contests and influence policy decisions by linking frames to specific ideas, norms, and values. Such communicative practices are particularly powerful not only because they are employed to pursue interests, but because they create them as well. Business actors seek to shape public discourses, which have an important influence over the way public debates proceed and the choices society makes (Clapp and Fuchs, 2009: 10).

Discursive power is also linked to the structural power of business. As some scholars point out, the privileged position of business actors is more likely to hold if its core values dominate public discourse (Guber and Bosso, 2007: 49). Should this change it could undermine the political legitimacy of business actors and in turn their capacity to shape future discourses. Indeed, for a business actor to effectively exercise discursive power it requires political legitimacy. While state actors mainly derive legitimacy from the formal authority associated with public office, business actors rely on the societies in which they operate viewing their behaviour as appropriate, as I will discuss below (Clapp and Fuchs, 2009: 10–11).

Business preferences

There is no doubt that business actors have the power to influence outcomes and this power has many dimensions. But power must be wielded, and business actors must mobilise to do so. The decision to mobilise and engage in policy contests is critical, because which actors are mobilised and which are not affects the balance of forces between actors, and ultimately therefore, policy outcomes (Schattschneider, 1960). Over recent decades, numerous studies have documented the growth in business mobilisation and the increased amount of resources business actors devote to national politics in the US (Shaffer, 1995; Vogel, 1989). Business and management scholars have identified a range of factors that affect business mobilisation. For example, some studies focus on the size of individual firms, seeing it as proxy for a firm's capacity to become politically engaged (Hillman *et al.*, 2004; Figueiredo and Richter, 2014). Others claim that political engagement is determined by the structure of the industry, such as the concentration of market share or the degree to which the industry is dependent on government (Shaffer, 1995: 501).

However, preferences are generally determined by interests. In policy contests, like those in the energy sector, business preferences will primarily be determined by the distributional effect of the policy. While some studies emphasise the collective interests of business, such as when all firms in an industry or a sector face a common threat, since the 1950s business unity in the US has been relatively uncommon (Smith, 2000). Instead, in most cases environmental regulations will have different costs and benefits for different industries and different firms within the same industry (Keohane *et al.*, 1998). In general, firms will tend to support regulations when they benefit from them and oppose them when they do not. This is the assumption of most studies in political science and related fields, which in essence, assume firm preferences are a function of their commercial interests and managers will be concerned about the distributive effect government regulation will have on firm profitability (Boddewyn and Brewer, 1994; Shaffer, 1995).

Further, because regulation impacts the competitive position of firms, which is a critical determinant of profitability, firms preferences are expected to reflect relative gains. Indeed, two firms may both stand to lose from the implementation of a regulation, but one firm might incur lower costs than a competitor, granting it relative gains. For instance, in the utility industry emissions standards on power plants will impose higher costs on plants that rely on coal than those that rely on gas. Hence, firms may support regulations that enable them to gain a competitive advantage over their rivals. The process of rent-seeking described by economists draws attention to the efforts of firms to gain a competitive advantage via changes in government policy rather than via competition in product markets (Tollison, 2012). In short, business actors make cost-benefit calculations on the likely distributional effect of different policies based on their competitive position in the market.

It should also be noted that business preferences are not fixed, and preferences can vary over time in the context of the policy cycle (Downie, 2014a; Vogel, 1989). For instance, in the agenda setting phase of policy development evidence from studies of environmental politics suggests that affected firms are more likely to pursue a strategy of opposition when regulations first emerge on the agenda, but as the debate matures, opposition is likely to decrease and become fragmented. Over time a tipping point can be reached where business tips from opposition to support (Vormedal, 2012; Meckling, 2015). In addition, policy feedbacks and path dependence can re-shape business interests and the balance between business interests by helping to determine what actions are possible, desirable, and legitimate (Stokes, 2015: Ch. 2; Layzer, 2012: Ch. 2).

Although commercial interests primarily drive preferences, they do not on their own predict the strategies business actors will pursue in policy contests. In other words, business actors' decision to support or oppose a given policy will not simply reflect commercial interests. As scholars in the sociological traditions of the literature point out, actors are embedded in institutional contexts that shape their decisions (DiMaggio and Powell, 1983). Hence, the actions of business actors are filtered through the institutions and networks to which they belong. For example, there is considerable research to indicate that national institutions matter and that business decisions to support or oppose policies will differ according to their domestic institutional contexts (Woll, 2008). In the case of the US, for example, the adversarial US legal system has often led corporations to take more antagonistic positions toward regulations than in Europe (Kagan, 1991, 2007). Such institutional factors help to explain the varieties of capitalism seen around the world, which are defended by firms that hope to gain a comparative institutional advantage (Hall and Soskice, 2001; Hall and Thelen, 2009). In addition, how business actors respond to their institutional contexts can also be affected by the unique history and culture of a firm or industry. For example, businesses that have experienced a history of losses with a particular technology are likely to institutionalise a negative view towards the future prospects of such technologies (Levy and Kolk, 2002: 208–281).

It is too simple then to speak of 'business preferences' as though business is a black box, with one set of consistent and coherent preferences. Business, like the state, is not a unitary actor and preferences will vary across industries and firms. While competition between businesses is what underpins competitive markets, these same business interests will regularly engage in fierce battles over the rules that govern those markets (Tienhaara *et al.*, 2012: 62). For example, business conflicts between national and international firms are common, especially around trade policy, just as they are between technology leaders and laggards in the same industry (Meckling and Hughes, 2015), or between companies that operate in different economics sectors, but within the same supply chain (Falkner, 2008: Ch. 2).

Business strategies

In order to understand the how and why of business behaviour it is necessary not only to understand the power and preferences of business actors, but also the mechanisms via which they attempt to influence policy outcomes. To be clear, policy contests are about the rules that govern markets. In the US rules are either legislated by congress and/or promulgated by government agencies, and the outcome of these contests can directly affect the competitive position of business.[1] Business strategies therefore refer to what business and management scholars might term non-market strategies, as opposed to market strategies, that is, strategies designed to influence the distributive effect of governance outcomes (Aggarwal, 2001; Figueiredo and Tiller, 2001; Mellahi *et al.*, 2016).

Business actors can employ a host of possible strategies, but their choice will be determined to a large extent by whether they choose to support or oppose a set of rules, or in some cases decide to take the middle ground and hedge their position. Baumgartner *et al.* (2009) argue that business actors seeking to change the status quo by advancing a new policy or rescinding an existing policy tend to be more active than actors defending existing policies. And, this will affect the strategies they choose to use, for instance, defenders of existing policy are likely to focus on the harm and costs of new policies, whereas advocates will seek to generate support for the proposed changes.

While much of the literature has historically considered business behaviour in terms of its support or opposition to particular policies, business can also hedge its position. This is particularly likely when there is strong pressure for policies to be implemented, which industries may be unable to stop, and hence they will look to shape rules that minimise their compliance costs rather than outright opposition (Meckling, 2015). These discussions in political science parallel similar discussions in business and management. Scholars in this tradition have long distinguished between buffering and bridging behaviour. Buffering in this context refers to proactive political activity by business actors to actively change government regulation via lobbying efforts, in contrast to bridging strategies that are a reactive form of behaviour and include efforts to track the development of new policies for the purposes of compliance or other objectives (Hillman *et al.*, 2004: 844).

In what follows I focus on three strategies that existing studies suggest are critical for actors in policy contests: mobilising coalitions (of multiple actors at multiple levels); framing and lobbying. The effectiveness of each of these strategies and the capacity of business actors to shape policy outcomes to a large extent will be a function of the resources business actors have at their disposal and their political legitimacy. First, resources include such things as financial resources, political, economic, and technical expertise, and the size of the support base (Doh *et al.*, 2012: 30; Sabatier, 1988: 143). As discussed above, financial resources, for example, are what give business actors instrumental power. The greater the resources a firm or a coalition of firms can amass, the greater its instrumental power. In the energy sector, there is every reason to believe that many business actors will possess considerable resources. After all, large corporations, especially incumbents in the fossil fuel industries, such as ExxonMobil, have annual revenues that are greater than the GDP of many nations.

Second, in broad terms a business actor has legitimacy when policy-makers, as well as its consumers and competitors, share a belief that its actions are 'desirable, proper, or appropriate within some socially constructed system of norms, values, beliefs and definitions' (Suchman cited in Bernstein, 2011: 24). In other words, a business actor has legitimacy when its behaviour is considered appropriate within the societies in which it operates (Collingwood, 2006; Clapp and Fuchs, 2009). As mentioned, legitimacy is important therefore because it is a source of both structural and discursive power. In the case of structural power, the more legitimate a business actor is considered by the communities with which it interacts, the greater its structural power is likely to be. This in turn means it is more likely to have institutional access to state actors (Mügge, 2011: 56–57). For example, business actors with structural power are more likely to be appointed to government advisory boards than those actors who do not, empowering those actors considered to be legitimate in policy contests (Bernstein, 2011: 20–21). In the case of discursive power, business actors' capacity to shape discourses, and in turn policies, is unlikely to be successful if they are not viewed as legitimate actors with the authority to participate in policy contests. Though the legitimacy of business actors is regularly called into question, in the energy sector at least, the actions of firms across the fossil fuels and renewable industries, are for the most part, considered appropriate among policymakers and the public. To be sure, energy corporations have played a large role framing the narratives around climate change (Schlichting, 2013).

Mobilising coalitions

Mobilising coalitions is a key strategy of business actors. In policy contests, the principal aim of coalition building is to influence the actions of policymakers and in turn shape policy outcomes. Coalitions enable this in two main ways. First, coalition building helps business actors to demonstrate the breadth and depth of support for their policy position. The simple rule is that the more actors one side mobilises the more likely they are to succeed in determining policy outcomes. To put it another way, the outcome of a policy contest will be determined by 'the scope of its contagion' (Schattschneider, 1960: 2). Second, coalition building allows business actors to leverage other strategies – framing and lobbying – because coalitions can help to pool resources and build legitimacy, two variables that determine the effectiveness of these business strategies. For example, to the extent that coalitions increase the amount of financial resources business actors have at their disposal, then coalition building will enhance the effectiveness of lobbying efforts. Similarly, when coalitions mobilise actors that enhance the legitimacy of their cause, they will in turn enhance the effectiveness of framing given that legitimacy is a source of discursive power.

So how do they mobilise coalitions? The short answer is that business actors often work to mobilise multiple actors at multiple levels. First, they can mobilise other business actors. Historically, building business coalitions has been a pervasive form of behaviour in US politics with industry, sectoral and economy wide business associations a common feature (Vogel, 1989). The rise of these types of coalitions occurred in the 1960s and 1970s when business actors' growing political engagement led to the establishment and reinvigoration of an array of business associations to represent their interests. For example, between 1960 and 1987 the number of industry specific associations with offices in Washington DC doubled (Vogel, 1989). Peak associations, such as the US Chamber of Commerce, went from an organisation with tens of thousands of members in the 1960s to one with hundreds of thousands in the 1990s, and today it remains one of the most important non-government actors in the US (Barley, 2010: 783–784).

Some industries are better organised than others to do this, often reflecting their experience in fighting long running regulatory battles. The pharmaceutical, chemical, and energy sectors are classic examples. In the energy sector, business actors have populated different industries with a variety of business coalitions. The most visible are industry associations, which are based around shared industrial preferences. For example, the oil and gas industries have the American Petroleum Institute, the coal industry has the American Coal Council, electric utilities have the Edison Electric Institute, solar and wind have their own associations and the list goes on, with most industries having many rather than one. Most of these associations and the firms that comprise them are also powerful actors in the peak organisations that represent business interests across the economy, notably the Business Roundtable, US Chamber of Commerce (COC) and the National Association of Manufacturers (NAM).

Further, formal trade associations are only one form of coalition. Since the 1970s, these industries, and sub-sets of firms in these industries, have also created ad hoc coalitions to maximise their political influence (Vogel, 1989). In general, ad hoc coalitions are informal coalitions that are temporary in nature, established to fight a single policy contest, and typically disband once the policy contest is over (Barley, 2010; Mahoney, 2007). While formal industry associations are easy to identify, ad hoc coalitions are not and often they are secretly created with names that obscure their agendas. This may explain why such coalitions are often overlooked in the literature, though as will be discussed in the following chapters, they appear to be playing a critical role in policy contests in the energy sector.

Business coalitions enable actors to pool existing resources within the business community rather than being forced to build them from the ground up, making coalition activity more economical and more powerful.

Studies of coalitions suggest that members will join coalitions to access additional resources, even though this may require actors to adjust their policy positions to be included in the coalition (Hula, 1999). At the same time, business actors also organise in ways to improve their legitimacy. For example, studies of firms in the biotechnology industry have shown that affiliations with a prominent business partner increase the value of the bio-technology firm (Podolny and Page, 1998: 64). As will be discussed in later chapters, the legitimacy and status-enhancing effects of enrolling more prominent business actors are likely to be important for new technology firms competing with incumbents in the energy sector.

Second, business actors can also build coalitions with other non-state actors. In the US there is a history of business actors and environmental NGOs, joining forces in policy contests over environmental regulation (Clapp and Meckling, 2013). Such coalitions have been labelled 'Baptists and bootleggers' coalitions because of their unlikely alliance, though as Desombre (1995) points out, on some issues they can have similar zones of agreement. For both business actors and environmental NGOs, these coali-tions can be effective because they provide legitimacy. Not-for-profit organisations, such as environmental NGOs, are typically seen as legiti-mate actors in public debates and can provide business actors with greater political legitimacy than if they were acting alone. Likewise, environmental NGOs can demonstrate to policymakers that the policy position they support has wider appeal outside their narrow membership bases. For both sets of actors then, mobilising such coalition sends a signal to policy-makers, especially elected officials, that there is a large set of interests sup-porting a particular position.

Third, business actors can also mobilise state actors to build their resources and legitimacy. Like business, the state is not a unitary actor, and different parts of the state will have different preferences. This has long been recognised in political science, especially in studies of bureau-cratic politics, which have shown that the preferences of government agen-cies and bureaucratic coalitions will be an important influence on the policy making process (Allison, 1971; Downie, 2013). Hence, business actors that are able to target and mobilise sections of government to their cause are more likely to be able to shape governance outcomes. For example, if oil and gas producers form an alliance with the US Department of Energy (DoE), or with members of the US Senate Committee on Energy and Natural Resources, this will not only build their legitimacy, but it will likely give them access to new resources, such as technical expertise.

Business actors also build coalitions across borders with other state and non-state actors. In other words, the coalitions that business builds are not restricted to the national level. While traditionally international relations has been dominated by state centred approaches, where the state is a unitary state, there is now an abundant literature, which can broadly be described as transnational relations, that shows the importance of cross-border coalitions (Keohane and Nye, 1972; Risse-Kappen, 1995; Slaughter, 2004). While much of this literature focuses on the role transnational coalitions play in determining global outcomes, such as in the field of climate change, they also influence national outcomes (Newell, 2000; Downie, 2014a). For example, transnational business coalitions are considered one of the main reasons national governments around the world have come to support emissions trading as policy tool to reduce greenhouse gas emissions (Meckling, 2011).

Whether at the national, international or transnational level, business actors are therefore enmeshed in wider networks of state and non-state actors. In many cases, business coalitions can play a key role working to loosely coordinate the activities of diverse sets of actors to shape outcomes. Scholars of regulation and governance, especially those that examine networks, often refer to these actors as 'brokers' or 'governing nodes' because they can act as a command centre bringing together members of different organisations to pool resources, share information and mediate conflicts to achieve a common purpose within a wider network (Burris *et al.*, 2005; Hadden, 2015; Hervé, 2014; Kauffman, 2017). For example, case studies have shown how the pharmaceutical giant Pfizer activated business coalitions first in the US, such as national industry associations, and then in ever-widening circles around the globe, enrolling business actors in Europe and Japan to re-write the global rules that govern intellectual property. In doing so, these networks were able to mobilise existing coalitions and draw them together to support their campaign (Sell, 2003; Sell and Prakash, 2004; Burris *et al.*, 2005).

Framing

A second strategy that business actors employ is framing. The concept of framing derives from sociological traditions in the literature that view ideas and discourses as important to explain policy stability and policy change (Schmidt, 2010; Snow, 2007). In this tradition, frames are defined as shared understandings of reality that structure how people behave. More precisely, frames refer to the 'specific metaphors, symbolic representations and cognitive clues used to render or cast behaviour and events in an evaluative mode and to suggest alternative modes of actions' (Zald, 1996: 262). Actors therefore use framing for three core tasks namely to: diagnose problems and attribute blame or causality; suggest solutions to problems and the strategies to achieve them; and to motivate supporters to address the problem, which typically involves emphasising the urgency of the issue (Snow and Benford, 1988).

Accordingly, business actors employ frames in policy contests to set agendas and draw attention to their concerns among policymakers and the wider community. Indeed, what emerges in many policy debates is framing contests where rival groups of actors compete to strategically frame debates. In other words, they engage in various rounds of framing and counter-framing as they try and replace an existing frame with their preferred frame (Sell and Prakash, 2004). Achieving the dominant frame is vital because of its capacity to shape discourses and, in turn, create and redefine preferences and influence policy decisions. Studies have shown that actors that succeed in establishing a frame that is consistent with their goals are likely to reap the greatest gains (Joachim, 2003; Odell and Sell, 2006; Snow and Benford, 1988; Guber and Bosso, 2007).

If the aim of framing is to set agendas, how do business actors engage in framing? In other words, what tactics can they employ? Though I will explore the tactics used by business actors in the following chapters, it is worth highlighting some of the ones identified in the literature. Of particular importance is framing the policy problem and the policy solution in ways that can be grafted onto existing frames or principles that have normative appeal in the community, including with policymakers. For example, policy proposals that are tied to positive value frames, such as free trade, freedom or security, have proven to be effective frames for business actors and NGOs because they have pre-existing normative appeal and hence are more difficult to counter (Sell and Prakash, 2004; Drahos, 2002; Guber and Bosso, 2007). In addition, such a tactic can be reinforced by the strategic use of information to reinforce frames. For instance, in contests with environmental NGOs, business actors have regularly turned to funding sympathetic policy think tanks to provide technical backing, such as economic modelling, which supports the claims being made (Layzer, 2007).

Such approaches reveal the different power dimensions that are often in play in framing contests. While business actors, for example, may have superior financial resources to shore up their preferred discursive frame directly, via public relations campaigns, or indirectly via the commissioning of favourable policy reports, such instrumental power can be countered. For instance, environmental NGOs often draw on their political legitimacy to make their case and/or to target the legitimacy of business actors. This was evident during the Deepwater Horizon oil spill in 2010 when BP's attempt to frame the public debate was limited by its perceived illegitimacy to continue to participate in discussions about policy solutions (Schultz *et al.*, 2012). More recently, the fossil fuel divestment movement has directly targeted not only the instrumental power of energy corporations by calling on shareholders to divest of their assets, but more importantly their discursive power, by challenging their social legitimacy to operate (Ayling and Gunningham, 2015; Ayling, 2017).

Lobbying

A third business strategy is lobbying. K Street, the famed walkway of lobbyists in Washington DC, is often used as a synonym for business lobbying in the US and political leaders regularly denounce its influence.

Yet many politicians retire to become lobbyists. In fact, in 2003 more than half of the retiring members of Congress returned to Washington DC as lobbyists (Barley, 2010: 790). And in 2016, $3.12 billion was spent on lobbying with business actors the biggest spenders. As a sector, energy contributed almost $300 million to that total, which is actually less than the $421 million it spent on lobbying in 2009, the first year of the Obama administration (CRP, 2016g).

With figures like this, it is no surprise that scholars of political science, and especially scholars of business actors, have focussed on lobbying in the US. Since the 1970s a wave of studies have examined who lobbies, the types of lobbying and its influence on policy outcomes. To some extent this was encouraged by changes to federal campaign finance laws that enabled scholars to access information about business contributions to political parties and candidates. As a result, studies of business finance came to dominate scholarship on business political activity (Vogel, 1996: 149; Figueiredo and Richter, 2014). In recent years, scholars have built on this work, for example, to study the array of lobbying tactics used by different interest groups (Baumgartner *et al.*, 2009; Grossman and Helpman, 2001). Surprisingly however, the lobbying activity of firms in the energy sector has received very little attention (Kim *et al.*, 2016; McFarland, 1984).

In general, lobbying refers to the various activities undertaken by business actors to educate and persuade lawmakers of the wisdom of their policy positions (Grossman and Helpman, 2001). A distinction is typically made between inside lobbying and outside lobbying activities. Inside lobbying is the most commonly employed activity and involves personal contact with policymakers, including White House staff, congressional members, and government agencies. Lobbyists devote considerable resources to building personal relationships with these government officials in order to build awareness about issues, propose, and amend legislation and regulations (Baumgartner *et al.*, 2009). In other words, inside tactics are the behind-closed-doors activities that are traditionally associated with lobbyists.

Inside lobbying tactics can also vary based on the nature of the policy contest. For example, in US legislative contests, business actors can lobby members of the House and the Senate, as well as key committee members, to delay or amend the legislation being advanced. Indeed, given the various veto points in legislative contests, there are multiple avenues for business actors to try and shape proposed bills (Klyza and Sousa, 2008). In regulatory contests where government agencies are tasked with writing regulations, business actors will also have opportunities to lobby given that government agencies in the US are required by law to take into account the views of stakeholders, such as via 'notice and comment' procedures (Yackee and Yackee, 2006). Surveys of business actors have found that they view such processes as an important avenue for influencing policy outcomes (Furlong and Kerwin, 2005).

In contrast, outside lobbying refers to the public campaigns that lobbyists organise via advertising, the media, and grassroots mobilisation, which aim to influence both the public and lawmakers (Grossman and Helpman, 2001; Baumgartner *et al.*, 2009). As Layzer (2007) argues, challengers to the status quo tend to use outside lobbying to convey the salience of an issue to lawmakers and create incentives for them to address it. Whereas, defenders of the status quo use these activities to prevent an issue from becoming salient by building confusion about the issue and pointing to the potential costs thereby making it less likely that elected officials will act.

The success of lobbying activities can depend both on the resources and legitimacy of business actors. Financial resources can give actors an advantage in gaining access to lawmakers, by enabling business to maintain a network of lobbyists devoted to cultivating personal relationships with policymakers. These resources can also be used to support public relations campaign. However, this does not mean that business actors will have it all their own way. Other non-state actors, such as public-interest groups, may be seen as more legitimate actors and their capacity to mobilise grassroots campaigns can prove an effective foil to business campaigns, as witnessed across multiple policy domains (Braithwaite and Drahos, 2000). Indeed determining the success of lobbying and its effectiveness remains 'extraordinarily challenging' and there is ongoing debate in the literature about the outcomes of empirical efforts to do so (Figueiredo and Richter, 2014: 168).

Conditioning factors

The aim of this chapter has been to consider the literature on business actors in order to examine their behaviour in the US energy sector, in particular to examine their preferences and strategies. However, considering how and why business behaves naturally leads to questions about their influence. In the policy contests in the US energy sector business actors were constantly in conflict, yet why did some actors win, and others lose? And what impact did these business battles have on the ultimate policy outcome? Studying business influence is a tricky task. Business influence cannot be directly measured and any conclusion about the impact of business actors, or that of any other actor, can be contested (Kraft and Kamieniecki, 2007). Business actors do not simply dictate their policy preferences to lawmakers who willingly adopt them. And even if they did, questions would still remain about why some business actors were successful and others were not. As discussed above, part of the answer lies in the behaviour of business; their power, their preferences and the strategies they choose to employ.

Yet business actors do not exist in a vacuum. Business behaviour is shaped by the environment in which it operates. Scholars in the social movement tradition refer to these factors as the political opportunity structure (Meyer, 2004; Meyer and Minkoff, 2004; Tarrow, 2005; McAdam, 2017). Broadly, that is, the 'dimensions of the political environment that provide incentives for people to undertake collective action by affecting their expectations for success or failure' (Tarrow, 1994: 85). The key insight of the political opportunity literature is that an actor's prospects for influence are context dependent. The challenge is determining those factors are likely to be important and those that are not. In this section, I follow the lead of Corell and Betsill (2008: 40) who argue that rather than constructing a single measure of political opportunity structures, it is most useful to think of political opportunity structures as clusters of variables and analyse how they have shaped opportunities for actor influence.

In the policy contests that took place in the US energy sector several conditioning factors that could be considered in such a cluster, for example, the role of international institutions, are likely to be less relevant (Tarrow, 2005). Accordingly, for the sake of parsimony I focus on two factors that may be critical to policy contests examined here: mobilisation of other non-state actors and the role of policymakers. First, the role of other non-state actors, such as environmental NGOs, will affect the influence of business actors. It is often assumed that when NGOs mobilise against industrial interests this will reduce their influence and vice versa. For instance, should fossil fuel industries mobilise against environmental NGOs this will likely limit their influence too. However, when business actors and environmental NGOs pursue separate agendas both may be able to have an influence without reducing the influence of the other. In other words, competition from other non-state actors is not necessarily a zero-sum game (Betsill, 2008).

In addition, just as competition from other non-state actors will impact on business actors, so too will cooperation. As discussed, business actors often look to cooperate with environmental NGOs in order to build legitimacy (Meckling, 2011). For business actors seeking to enhance their political legitimacy, cooperating with environmental NGOs via formal and informal coalitions, can prove particularly useful to boost their discursive power and in turn their capacity to frame debates and set agendas (Raustiala, 1997). The added benefit of such cooperation is that it demonstrates that a particular policy position has broad support across interest groups that are typically opposed to one another, which is one of the reasons Baptist and bootlegger coalitions emerge (Desombre, 1995).

Second, the influence of business actors will also be conditioned by the role of policymakers who have the capacity to significantly affect the influence of business actors by creating and limiting their opportunities to shape outcomes. Policymakers by definition are key actors in policy contests. But what drives policymakers to make the decisions they do? Existing research suggests that the answer is beliefs and political incentives. On the one hand, the beliefs or ideas that policymakers have about the nature of a particular problem, like trade, energy or the environment, and the workings of particular policies, like emissions trading, will impact their preferences (Layzer, 2012; Sabatier, 1988). On the other, in democracies elected officials are also driven by their political incentives, namely their desire for re-election (Putnam, 1988). Further, the political incentives of policymakers will often trump ideas about particular policy issues, especially when these come into conflict with what voters want (Harrison and Sundstrom, 2007). Smith (2000) argues that the most important factor explaining whether business succeeds or fails is public opinion. When the public opposes the preferences of business, policymakers tend to follow the public. For example, should proposals to reduce emissions from the energy sector have greater public support than maintaining the status quo, policymakers are likely to act.

Towards an analytical framework

Business is an influential actor in American politics and its behaviour has been well-studied in the social sciences. However, key questions remain about how and why business actors are shaping energy policy contests in the US. Indeed, very little work has focussed exclusively on business actors in the US energy sector, especially in the context of the recent transformations that have been witnessed across the oil, gas, coal, utility, wind, and solar industries. Drawing on the insights of scholars in political science, business and management, and regulation and governance, the aim of this chapter has been to set out an analytical basis for examining business actors in the US energy sector.

In examining business behaviour, the first issue to consider is the preferences of business actors. In other words, when business actors mobilise in a policy contest why do they adopt the positions they do? The short answer is that business preferences are generally a function of their commercial interests. Business actors will support policies that improve their competitive position in the market and oppose those that do not. Though if the distributive effects on firms in the same industry are different, then it is anticipated that preferences will differ to reflect relative gains. For example, as noted above, in the utility industry emissions standards on power plants will impose higher costs on plants that rely on coal than those that rely on gas. However, commercial interests on their own will not predict the strategies business actors employ, these will be affected by the institutional contexts in which business actors are embedded. For example, national institutions can impact on the types of strategies business pursue, and how business respond to the institutional context will also be influenced by the unique history and culture of a firm or industry. Nevertheless, the behaviour of business actors in the energy sector is likely to be driven by commercial interests.

Understanding business preferences may help to explain why business actors do what they do, but in order to examine how business actors seek to support or oppose the status quo, it is necessary to consider their strategies. It can be expected that business actors will primarily rely on three strategies: mobilising coalitions (of multiple actors at multiple levels), lobbying, and framing. Business can mobilise other business actors via formal trade associations or via ad hoc coalitions. They can also be expected to build coalitions with other non-state actors, such as environmental NGOs or consumer groups, and to mobilise state actors in support of their cause. Such coalitions are not restricted to the national level and existing studies have shown the power of transnational business coalitions. In many cases, business actors are enmeshed in diverse networks of actors and business coalitions can play a key role coordinating the activities of such networks of actors. Business will also engage in inside lobbying, building personal relationships with lawmakers, and outside lobbying by mounting public campaigns to influence not only policymakers, but public opinion too. And they will seek to frame policy contests to set agendas and draw attention to their concerns. This may include tactics, such as framing a policy problem or solution in ways that can be grafted onto existing frames that have normative appeal in the community, or by reinforcing the frame with the strategic use of information.

Critically however, the effectiveness of these strategies will be a function of the resources of business actors and their political legitimacy. Resources, especially financial resources, provide business actors with the instrumental power to shape governance outcomes. This is the very reason business mobilises coalitions, to pool resources that can be used, for example, to fund more extensive lobbying efforts. The effectiveness of business strategies also depends on the legitimacy of business actors, which reflects whether their behaviour is considered appropriate among the communities they interact with. It is this legitimacy that provides business with the structural power to, among other things, gain institutional access to key decision makers. It also provides the discursive power to frame policy contests and set agendas and to shape rules more effectively when they are viewed by policymakers as credible stakeholders in the process.

It is expected that this cluster of variables will help to explain much of the business behaviour witnessed in the US energy sector. However, the ongoing challenge is attempting to understand the influence of these actors and the impact they have had on governance outcomes. To some extent, this can be done by examining the variables above, that is, by considering the preferences of business actors and the strategies they employ. In short, to study their actions. This is the approach taken here. Yet the framework presented in this chapter suggests that this is not enough. Business behaviour is also conditioned by its political opportunity structure. Accordingly, it is necessary to identify those factors that are likely to shape the decisions and actions of business. In the policy contests examined here two are likely to be crucial: mobilisation of other non-state actors and the role of policymakers. To varying degrees the role played by other non-state actors, such as environmental NGOs, or by policymakers, can directly create and limit opportunities for business actors to determine policy outcomes.

By setting out an analytical basis for examining business actors in the US energy sector that combines the cluster of variables most likely to determine business behaviour, namely preferences and strategies, with the relevant conditioning factors, this approach goes beyond those more narrowly focussed on business preferences, or for example, specific strategies, such as lobbying. Of course, it is by no means comprehensive, and there are likely to be additional factors that will restrict or enhance the influence of business actors. In the following chapters this framework will be used to guide the empirical research, particularly to understand the how and why of business behaviour, but it cannot claim to provide a definitive explanation for the outcomes in each of the policy contests or definitively assess the influence of business actors. Instead, via careful exploration through empirical and context specific research the aim is to identify the principal mechanisms business actors in the energy sector use and to approximate their likely effect. In other words, to reconstruct the behaviour of business actors via the process tracing that good historians and journalists do.

The next chapter starts by examining business battles in the oil and gas industry.

Note

1 Studies of interest groups in political science, especially those in the American tradition, tend to distinguish between the legislative acts of congress and the rules of government agencies (Furlong and Kerwin, 2005).

3   Exporting to the world

Policy contests in the oil and gas industries

Introduction

> America's emergence as a global energy leader has fundamentally reordered the world's energy markets.[1]

Jack Gerard, CEO of the American Petroleum Institute (API) delivered these words in January 2015 at the fifth annual launch of the State of American Energy report. The triumphant tone reflected a set of facts that would have seemed unimaginable less than a decade ago. The US is now the largest producer of oil and gas in the world and is on track for energy independence. Since President Nixon proposed Project Independence in 1973, with the aim that the US would meet its own energy needs by the end of the decade, the nation has been on a failed pursuit for energy independence (Solomon and Krishna, 2011: 7427).

It appears that goal is now closer than ever. The reason is the 'shale revolution'. The technology breakthroughs that have allowed producers to access the vast onshore oil and gas reserves locked in impermeable rocks has transformed US energy markets and, as Jack Gerrard proclaimed, global markets as well. In 2008, the US was still processing approvals for new LNG import terminals to meet gas demand, yet within four years it had overtaken Russia to become the world's largest gas producer (IEA, 2013a: 199). It is the same for oil. Between 2003 and 2013, US oil production was 30 per cent higher. Within a decade the US has gone from being the largest oil importer to the largest oil producer surpassing both Russia and Saudi Arabia (IEA, 2014a: 173).

Just as it did in the twentieth century, the US oil and gas industries look set to have a critical impact on the future of the twenty-first century. After all, natural gas is the largest energy source in the US and it remains a significant source of global greenhouse gas emissions (EIA, 2018a). Recent estimates indicate a third of the world's oil reserves and half of the world's gas reserves will need to stay in the ground to stay within the 2°C guardrail (McGlade and Ekins, 2015). And, importantly, the US is home to not only three of the largest oil and gas companies in the world – ExxonMobil,

Chevron, and ConocoPhillips – but when combined with the thousands of independent producers it also has the largest oil and gas industry in the world (EIA, 2018g).

Accordingly, business actors in the US energy sector will be critical to determining the speed and scale at which an energy transition can be achieved. For example, political resistance from the oil and gas industries could derail government efforts to mitigate greenhouse gas emissions. Yet, as discussed in the last chapter, very little work has focussed exclusively on business actors in these industries, especially in contemporary policy contests in the US. This chapter begins to redress this blind spot in the literature. The recent policy contests triggered by the shale revolution provide an excellent window through which to consider how and why business actors are shaping governance outcomes in these industries. The focus is on two contests: gas exports and oil exports. Specifically, the push by gas producers to lift export restrictions, which was opposed by petrochemical manufacturers, and the push by oil producers to overturn export restrictions on crude oil, which was resisted by the refining industry. The contests are messy and complex, but the detail matters because it is critical to understanding why business actors behave the way they do, that is, their preferences. And, how they seek to exercise influence over policy processes and shape outcomes. The next section provides an overview of the US oil and gas industries, including the business actors. The following sections examine the preferences and strategies of these actors in the two contests.

Overview of the US oil and gas industries

The US is now the largest oil and gas producer in the world and most projections expect this to continue in the coming years. This has significant ramifications for energy markets and efforts to achieve a transition away from fossil fuels. First, oil is the most significant energy source in the world (IEA, 2016a). In the US oil represents around a third of the primary energy supply (EIA, 2018h). While for many decades US demand for oil has outstripped supply, this is now changing with the shale revolution driving an unprecedented increase in domestic production. In 2016, the US was the largest producer in the world of petroleum and other liquid fuels, producing 14.8 million barrels per day – more than Saudi Arabia and Russia (EIA, 2017f). Crude oil production comprised almost 8.5 million barrels per day, almost half of which was produced directly from shale oil resources (EIA, 2017f). This historic shift in US production is projected to continue. The EIA estimates that US crude oil production will surpass the record 9.6 million barrels per day and plateau between 11.5 to 11.9 million barrels per day. Under a high oil price scenario the EIA estimates that by 2050 crude oil production could reach 19 million barrels per day, an extraordinary amount considering that total

US crude oil production was less than six million barrels per day in 2000 (EIA, 2018a).

Second, while gas comprises around a fifth of the world's primary energy supply, it remains a significant energy source in the US and around the world for electricity, transport, and industry (IEA, 2016a: 14). As a result of the shale revolution, US gas production has also grown dramatically. In 2017 the US was the largest producer of natural gas in the world, with around 90.9 billion cubic feet per day (EIA, 2018b). And the EIA expects natural gas production in the US to continue increasing reaching around 100 billion cubic feet per day by 2050, and reaching close to 140 billion cubic feet per day under a high price scenario. In other words, close to double US natural gas production today. Again, the growth in shale gas helps to drive these projections (EIA, 2018b).

The increase in US oil and gas production has widespread implications domestically and internationally. In the US, surging oil production and limits on oil exports has led to a renaissance in the US refining industry, as low domestic oil prices have increased the international competitiveness of refiners that produce gasoline and aviation fuel, among other products. For example, large US refiners, such as Motiva and Marathon, have added new refining capacity in recent years to take advantage of new domestic production from the shale resources (IEA, 2014a). Likewise, the growth in natural gas production and the subsequent historically low domestic gas prices have been a boon for some parts of US industry, most notably petrochemical manufacturers, which rely on natural gas as a feedstock. Globally, increases in production have meant that the US has reduced its reliance on import markets and is increasingly looking to export oil and gas. For example, much of the natural gas import capacity built in the early 2000s is now sitting idle or is being converted into export facilities (IEA, 2014a). The EIA projects that natural gas production will increase modestly until 2020 (EIA, 2018a: 16). After five additional natural gas export facilities come online in 2021, LNG export capacity is projected to increase as demand in Asia grows and US gas prices remain competitive (EIA, 2018b). While the US remains an importer of oil, its dependence on oil imports is declining and there is an expectation that US oil exports could grow, especially under a high oil price scenario now that Congress has removed restrictions on oil exports; one of the two policy contests discussed in this chapter (EIA, 2017b).

However, increasing US oil and gas production is inconsistent with limiting global average temperatures to 2°C – the principal aim of the Paris Agreement. The IEA projects that under its 450 scenario, which aims to limit temperatures to 2°C, global oil demand would peak by 2020 and then steadily decline, with the US experiencing the largest decline among OECD countries to 2040. One of the major drivers is reduced demand from passenger vehicles as oil use is replaced by a greater uptake in electric vehicles and biofuels (IEA, 2016b: 111–112). While gas demand continues to increase until the mid-2020s, it too levels off, because as the IEA 450 scenario makes clear 'even gas is too carbon intensive for long-term growth in a decarbonising energy system'. In the US gas generation in the electricity sector peaks around 2025 and then the decline accelerates out to 2040 (IEA, 2016b: 167). Should this scenario become a reality, and it remains a big if, lower demand could result in surplus supply and stranded assets, according to the IEA's forecasts. The risk of stranded assets – capital investment into assets that is not recovered – is more acute in the oil industry than the gas industry. This reflects the capital intensive nature of oil production, especially conventional oil production coupled with the projected decrease in demand under the 450 scenario, which is greater in the oil industry than the gas industry, including in the US (IEA, 2016b: 154).

So who are the business actors in these industries? The US oil and gas industries are dominated by the so-called 'supermajors', which trace their history back to Standard Oil and the Seven Sisters (Sampson, 1975). In the period of the policy contests in the US Shell is the largest oil and gas corporation operating in the US based on annual revenues, which totalled $421 billion in 2014 (Shell, 2015), followed by ExxonMobil, BP, Total, and Chevron, among others, including ConocoPhillips, which has the largest share of domestic oil production (Witter, 2015a). In addition, there is a strong tradition of independent producers in the US, defined as those companies, which do not have more than $5 million in retail sales in a year (IPAA, 2013). There are many thousands of such producers that have played a crucial role in the development of the industry, as epitomised by George Mitchell, an independent oil producer in Texas, who first proved the fracking technology financially viable in the late 1990s. The vast majority of oil and gas production takes place in Texas, Oklahoma, Louisiana, and Colorado (Witter, 2015a). The API is the most prominent industry association. Other associations, such as the Independent Petroleum Association of America (IPAA), which represents independent producers, and the American Natural Gas Alliance (ANGA) often work closely with the API.[2]

Many of the large oil and gas corporations are fully integrated, which means that not only do they produce oil and gas, but they also refine crude oil into products such as gasoline and diesel fuel oil. In addition, many of these firms also have a chemical segment and hence manufacture and sell petrochemical products, such as propylene and benzene. For example, ExxonMobil, Chevron, and Shell are large refiners and large petrochemical manufacturers. Though they are not the only ones. Other corporations that dominate the refining industry, such as Valero Energy, only operate refineries (Witter, 2015b; IEA, 2014a). Similarly, DowChemical, which dominates the petrochemical manufacturing industry in the US, does not have any upstream oil and gas operations (Blau, 2015). Integrated companies, like ExxonMobil, therefore, can have different commercial interests to those that simply operate refineries, or only manufacture petrochemicals, though as will be discussed, both are directly affected by the rules governing exports markets, albeit in different ways. In the US, the majority of refining is undertaken in Texas, California, and Louisiana (Witter, 2015b). The principal industry associations for refiners and petrochemical manufacturers' are the American Fuel and Petrochemical Manufacturers (AFPM) and the American Chemistry Council. The API also represents many of these corporations.

The oil and gas industries have experienced waves of restructuring. In the mid-1980s the plunging oil price and falling revenues, forced the largest players to reduce costs, increase efficiency and establish the scale required to undertake multi-billion dollar offshore projects that were appearing on the industry's horizon, which meant a series of mergers and acquisitions (Parra, 2004: 324; Yergin, 2008: 765). Fast forward 30 years and further restructuring occurred following the fall in the oil price. Between mid-2014 and early-2015, the price of oil dropped from over $100 per barrel to below $50 (IEA, 2015a: 20). And, like then, mergers followed, such as Shell agreeing to takeover BG Group, one of the largest gas producers in the world, for $70 billion dollars in April 2015 (Kollewe and Farrell, 2015).

The policy contest over gas exports: 'we have broken a lot of glass'[3]

The sharp increase in US gas production led to a contest between gas producers and petrochemical manufacturers over the rules governing exports. The reason was simple, since the shale revolution a gap had emerged between the domestic price for gas and the spot price on international markets. In the gas industry there are three regional markets. In North America prices are determined at hubs, such as Henry Hub, which reflect local supply and demand, whereas in the Asia-Pacific region and Europe the price of gas has traditionally reflected long-term contracts linked to the price of oil. As a result of the shale boom, the Henry Hub price in the US fell from as high as $12 per million Btu in 2008 to less than $2 per million Btu in 2012, while international prices remained as high as $17 per million Btu (EIA, 2016d, 2011). Gas producers were eager to take advantage and sell LNG to Asia and Europe (API, 2014b). As one prospective exporter argued:

> It makes sense to export gas given the excess supply and the spreads between the Henry Hub price, which was $2, and the price in Southeast Asia, such as in Japan, which was $14.
>
> (Interview 40)

While there was no legislative ban on natural gas exports, there were regulatory restrictions, which meant any changes required the support of the

White House rather than Congress. *The Natural Gas Act of 1938* and *the Energy Policy Act of 1992* in effect restrict the export of natural gas to countries with which the US has signed a free trade agreement (FTA). For exports to non-FTA countries, which include China, Japan, and the EU, the Department of Energy (DoE) can provide approval should it deem it to be in the 'public interest' (DoE, 2015b). However, delays in the approval process and controversy over what was meant by the 'public interest' led gas producers to lobby for the process to be reformed to allow export terminals to be approved more quickly. They were joined by Republicans and Democrats in Congress who urged the White House to act, though the Democrats were split with many raising concerns about the impact large-scale exports of natural gas would have on consumers and business. Ultimately, the industry was successful, and the process was expedited. By 2016 the administration began to approve new export terminals for natural gas.

The policy contest commenced in 2011. According to oil and gas executives the API and ANGA, the two principal industry associations with significant financial resources 'took the lead' and coordinated the campaign (Interview 40). They were strongly supported by producers, such as ExxonMobil, Chevron, and ConocoPhillips, along with prospective gas exporters, including Cheniere Energy, Dominion Resources and Sempra Energy, which all supported easing exports restrictions (Interviews 19 and 40; Ebinger and Avasarala, 2013).

Much of the debate centred on the perceived economic costs and benefits. The API and ANGA claimed that permitting exports 'will create thousands of US jobs, generate billions of dollars in revenue, improve our trade deficit and spur major investment in infrastructure, which will strengthen our energy security' (Gerard, 2013). To inform the debate, DoE commissioned two studies to assess the economics of gas exports. The first by the EIA, an independent agency under the DoE, and the second by NERA Economic Consulting, were released in late 2012. While both reports generally supported increased natural gas exports, the report by NERA, which concluded that lifting export restrictions will generate 'net economic benefits' set off a maelstrom in the business community (NERA, 2012). In fact, DoE received over 188,000 initial comments to the reports (Smith, 2015).

By early January 2013, all of the major US business associations were involved. On 9 January Thomas Donohue, president and CEO of the COC, came out in support of natural gas exports claiming they will bring jobs and growth to the US (US Chamber of Commerce, 2013). The COC was joined four days later by the NAM, which made a similar case citing the recent economic studies commissioned by DoE (Dempsey, 2013). It was not just economics that was used to frame the case; industry tied the case of gas exports to the principle of free trade. As Ross Eisenberg Vice President at NAM stated at the time, the association was founded in 1895 on the principles of free trade, which is why it 'fundamentally supports free trade and open markets', including for natural gas (Eisenberg, 2013a). They were not the only ones, throughout the contest, the leading actors, such as Marty Durbin, CEO of ANGA, consistently relied on their discursive power to frame the contest as a question of free trade,

> Just as the trade of any commodity promotes domestic jobs and economic growth, so too will the trade of natural gas.
>
> (Durbin, 2015)

However, the decision of the COC and NAM to support natural gas exports, ahead of a series of Congressional hearings on the issue, triggered a 'food fight' in the business community (Interview 20). The resistance to the oil and gas industry was led by Dow Chemical, one of the largest chemical and plastics manufacturers in the world, which relied on gas as a feedstock for its manufacturing plants. Dow argued that historically low gas prices had spurred a 'manufacturing renaissance' in the US with manufacturers planning to invest billions of dollars in new production facilities for the first time in over a decade. Yet Dow also argued that the potential price increases and volatility that unrestricted exports could generate threatened this renaissance (Dow Chemical Company, 2013). Dow's resistance had the support of a number of high profile Senators, including the incoming chairman of the Senate Energy and Natural Resources Committee, Democrat Senator Wyden, who had written to Energy Secretary Chu, criticising the DoE's analysis of the economic impacts of lifting restrictions (Anon., 2013).

In the first weeks of January 2013, Dow withdrew its membership from the NAM over its support for natural gas exports and almost did the same from the American Chemistry Council, until it decided to temper its position on exports (Krauss and Schwartz, 2013). In response, along with Alcoa, Eastman Chemical, and Huntsmen Chemical, Dow Chemical launched a new group called 'America's Energy Advantage', which called on the DoE to disregard the economic studies (Interviews 10 and 20). According to executives involved 'Dow was the ringleader' (Interviews 19, 20, 35). For Dow and America's Energy Advantage, the aim was to 'slow down' the approval process for new export terminals' because 'they were enjoying the bounty of natural gas' (Interviews 10 and 20). The longer gas exports were restricted, in their view, the longer domestic gas prices would remain low and their profit margins high (Interviews 10, 20 and 42). Polling released by the opponents showed that around 80 per cent of Americans were on their side, largely because of concern around gas prices (Markey, 2013).

The splits in the business community were 'nasty' (Interview 20). As one insider quipped, 'we have broken a lot of glass on the hill' (Interview 42). By mid-2013, the contest had spilled onto the pages of the *New York Times* with the CEO of Cheniere Energy, labelling Dow Chemical CEO, Andrew Liveris, 'a hypocrite and a self-serving person' (Krauss and

Schwartz, 2013). Similarly, the API derided the 'short-sighted efforts by a few industrial users to restrict exports in an apparent attempt to control prices' (Geman, 2013).

Prospective gas exporters, such as Cheniere Energy, Dominion Resources, and Sempra Energy, did not establish a similar ad hoc coalition to America's Energy Advantage, though along with gas producers, these firms worked to leverage transnational networks. In February 2013, at the US–Japan Summit in Washington DC, Japanese Prime Minister Shinzo Abe urged President Obama to approve natural gas exports to Japan as quickly as possible. Following the Fukushima disaster, Japan had shut down its nuclear power industry and it continued to search for alternative energy supplies (Ministry of Foreign Affairs Japan, 2013). While the oil and gas industry did not directly play a hand in this lobbying, respondents claimed that 'Japan was a tremendous ally and advocate for exports' telling the US administration that 'they would rather be taking US gas than gas from Russia or Iran' (Interview 40). In doing so, these state actors helped to provide gas producers with greater legitimacy, by demonstrating to policy-makers that support for the campaign was not simply confined to narrow commercial interests. As one described it,

> Here in DC a single company has trouble making an argument on their own, you have to demonstrate the size of your position and broad breadth of your support.
>
> (Interview 40)

By early 2014 the oil and gas industry was playing a direct role garnering international support. In March, API and ANGA among others, used their financial resources to help fund LNG Allies, an alliance between US producers and Eastern European governments that wanted access to US natural gas exports (Juliano, 2014). Drawing on geopolitical arguments, such as the need for the US to help its allies diversify from Russian gas, especially in the wake of the crisis in the Ukraine, the industry pressed Congress to act (API, 2014a). Within weeks, the ambassadors for Hungary, the Czech Republic, the Slovak Republic, and Poland were lobbying Congress to expedite the approval process for exports, with the vocal support of the API and the Republican leadership, including House Speaker John Boehner as well as a number of Democrat senators, such as Senator Heitkamp who had a long association with the industry having been a director of a natural gas company prior to her election to the Senate (Ryan, 2014; Lipton and Krauss, 2015). LNG Allies also helped organise the Ambassador of the Czech Republic, Petr Gandalovič, to testify before Congress in support of gas exports (Interview 16) (US House of Representatives, 2015). And they were not the only group. Other industry associations worked to enlist their European counterparts to make the case to the Obama administration. As one industry insider explained:

We have been encouraging Business Europe to let their governments know that it wants our government to export oil and gas ... we have governments coming through these hallways all the time, particularly the Polish.

(Interview 14)

Ultimately, the industry campaign succeeded with the DoE expediting the approval process for new terminals and by 2016 the first gas export terminal had been approved – Cheniere Energy's Sabine Pass. Further terminals have now also been approved and are under construction (EIA, 2016c). To some extent, the resistance from the chemical manufacturers led by Dow Chemical has been successful. By slowing down the approval process for export terminals, they were able to profit from a set of market conditions that were never going to last i.e. a domestic price almost nine times lower than international prices for gas. In the absence of such resistance it is likely that producers and exporters would have been able to access international markets and international prices in Europe and Asia much sooner.

Why and how did business actors shape the contest over gas exports?

Preferences

In the policy contest over gas exports, the evidence shows that business actors formed preferences based on their commercial interests. Gas producers, such as ExxonMobil, Shell, and Chevron, and prospective exporters, such as Cheniere Energy, Dominion Resources, and Sempra Energy, uniformly supported easing restrictions on gas exports for the simple reason that for producers it would allow them to access higher prices on international markets and for exporters it would open up new markets in Asia and Europe. There was no evidence of outliers, support for easing restrictions was uniform across the gas industry. As expected therefore, there were no signs of intra-industry conflict, and the principal industry associations, such as the API, ANGA, and the IPAA, all supported gas exports.

That said, not all business actors supported gas exports. Yet, here too, the reasons reflected the different commercial interests of their industry, which in turn led to inter-industry conflicts. For example, in the petrochemical industry, petrochemical manufacturers, led by Dow Chemical, opposed unrestricted gas exports because gas was a feedstock and, in their view, exports would result in higher domestic gas prices, higher costs of production, and therefore lower profits. As the CEO of Huntsmen Chemicals argued, 'if all the proposed gas export projects were built with reckless abandonment ... then the US price of gas would skyrocket' (Crooks,

2013). However, aware that they were unlikely to restrict gas exports forever, these firms hedged their position by seeking to delay the removal of the restrictions for as long as possible. For Dow and America's Energy Advantage, the aim was to 'slow down' the approval process for new export terminals' because 'they were enjoying the bounty of natural gas' (Interviews 10 and 20). The longer gas exports were restricted, in their view, the longer domestic gas prices would remain low and their profit margins high (Interviews 10, 20 and 42). As will be discussed, this type of hedging behaviour appears especially prevalent in the US energy sector.

Strategies

To realise their preferences business actors employed a number of strategies in their attempts to shape the policy contest (see Table 3.1). As discussed in Chapter 2, the effectiveness of these strategies to a large extent is a function of the resources business actors have at their disposal, such as financial resources, which give business actors instrumental power, and their political legitimacy, which can be a source of both structural and discursive power. First, actors on both sides of the contest – gas producers and petrochemical manufacturers – mobilised coalitions, which enabled them to pool resources and organise collective strategies. The API and ANGA, the two principal industry associations, acted as the command centre for the campaign for gas exports (Interviews 8 and 20). In doing so, they relied on their structural power to tie together coalitions across the wider business community, including the COC and the NAM, two of the largest business coalitions in the country.

Further, business actors did not just mobilise traditional associations, they also established ad hoc coalitions, which are characterised by their temporary nature and exclusive focus on a single policy contest, in this case gas exports. In the petrochemical industry, Dow Chemical helped to establish America's Energy Advantage, an ad hoc coalition of petrochemical manufacturers opposed to lifting the restrictions on exports. In part, this resulted from the fact that those opposed to gas exports could no longer rely on their traditional industry association, the NAM, to support their campaign, because it had come out in support of exports and hence, they were left to create alternative organisational forms. As will become apparent in the following chapters, these types of coalitions provide actors with significant flexibility to mobilise in support or opposition to specific regulations.

Business actors also built transnational networks with state allies. A case in point was LNG Allies, which gas producers established to promote exports by enlisting Eastern European governments to bring pressure to bear on federal legislators (Lipton and Krauss, 2015). As noted, they were able to enrol the ambassadors of Hungary, the Czech Republic, the Slovak Republic, and Poland to petition Congress. At the same time, other industry associations, such as the COC, sought to activate business actors

Table 3.1 Summary of the policy contest over gas exports

	Gas producers	Petrochemical manufacturers
General industry preferences	Support	Oppose/hedge
Key coalitions mobilised in the energy sector	API, ANGA, IPAA	America's Energy Advantage
Key business coalitions mobilised	COC and NAM	–
Other important actors in the networks	LNG Allies (includes Poland, Hungary, Czech Republic and Slovak Republic)	–
Framing strategy	Framed around free trade	Framed around manufacturing renaissance
Lobbying	Oil and gas industry spent more than $140 million per year on lobbying	Dow Chemical spent around $9 million per year on lobbying
Outcome	Restrictions on gas exports eased	

in Europe to make the same case (Interview 14). The effect was to create a network of transnational state and non-state actors in support of natural gas exports.

Second, business actors worked to frame the debate and set the agenda. In order to garner support for easing restrictions on gas exports, producers used their discursive power and framed the solution as free trade. This was effective because as an existing frame, free trade had normative appeal among business actors and policymakers whose support would be useful. Indeed, the free trade frame was used to draw in an ever-widening circle of actors to the cause. For example, NAM's support for natural gas exports was tied to its longstanding support for free trade. As noted above, the Vice President of NAM claimed that one of the reasons it supported gas exports was that NAM 'fundamentally supports free trade' and had so since 1895 (Eisenberg, 2013a).[4] In a series of testimonies before Congress NAM consistently made this link, arguing that 'LNG exports should be governed by principles of free trade and open markets' (Interview 35). As the debate progressed, gas exporters also used framing to link issues. For example, free trade in gas exports was linked to the crisis in the Ukraine, with gas producers arguing that the US should help its allies in Europe diversify from Russian gas.

Third, firms in the oil and gas industries leveraged their financial resources to undertake a vast lobbying campaign. Given that most of the top oil and gas corporations had annual revenues in the hundreds of billions of dollars, it is no surprise that between 2010 and 2015 the oil and gas industry as a whole spent more than $140 million per year on lobbying (CRP, 2015c). And gas exports was one of the main issues. Industry associations, especially the API and ANGA, 'had a huge responsibility lobbying government' (Interviews 7 and 35). For example, between 2011 and 2013, ANGA spent $7.5 million dollars lobbying on behalf of major gas producers (CRP, 2016a). In return, petrochemical manufacturers spent millions too, for instance, Dow Chemical spent $27.5 million on lobbying over the same period (CRP, 2015b). Much of this was directed at inside lobbying. In interviews with industry representatives, almost all described lobbying members of the House and the Senate to establish Congressional hearings and or meeting with White House officials to push for an end to export restrictions. Some business actors sought out senior advisors in the Obama administration to lead the effort. For example, Cheniere Energy hired Heather Zichal, President Obama's deputy assistant secretary for energy and climate change (Halperin, 2014).

Finally, what influence did business actors have on the policy contest? As discussed in Chapter 2, assessing business influence is no easy task and any conclusion about the impact of a particular actor, or set of actors, can be contested. However, to an extent business influence can be assessed by examining business preferences and strategies. For example, the actions taken by the Obama administration to accelerate the approval process for gas export terminals and ultimately begin approving export facilities is unlikely to have occurred without the campaign from gas producers. Similarly, should there have been no resistance from petrochemical manufacturers, it is also likely that this would have happened sooner.

Further, as outlined in the previous chapter, the influence of business actors on the policy contest is likely conditioned by the mobilisation of other non-state actors and the role of policymakers. While other actors, such as environmental NGOs, do not appear to have played a significant role in this contest, the positions of policymakers did create and limit opportunities for business actors to shape policy outcomes. First, the cautious approach of the Obama administration to unfettered gas exports likely reflected the belief of many in the Democratic Party that increasing exports would have negative environmental consequences, especially in the context of climate change. Perhaps more importantly, it also reflected their domestic political incentives, with many Democratic senators voicing concerns about the potential economic impact on domestic consumers and businesses, which was reinforced by opinion polling showing concern within the community about a rise in gas prices (Goode, 2014; Markey, 2013). This position created opportunities for business actors, such as petrochemical manufacturers, to exploit these concerns about the economic impact of permitting gas exports via their lobbying activities. At the same time, it constrained the capacity of gas producers, who worked to allay such concerns, including among Republicans that supported gas exports.

The policy contest over oil exports: 'oil producers engaged in a textbook public campaign'[5]

Since the oil shocks of the 1970s, which saw oil prices soar and queues for gasoline across the country, the US had largely restricted the export of crude oil.[6] However, with oil production almost doubling between 2011 and 2015 producers initiated calls for restrictions on oil exports to be removed (EIA, 2015b). The reason was a commercial one. The spike in US production resulted in a price spread between the domestic price (the WTI) and the international price (Brent). Between 2011 and 2014 the price of the WTI averaged $14 per barrel lower than the Brent (GAO, 2014: 7). Yet export restrictions meant oil producers could not access international markets. While many Republicans in Congress supported lifting the restrictions, the majority of Democrats did not, and neither did the White House. As a result, a contest ensued as oil producers attempted to overturn a ban on crude oil exports that had been in place for 40 years. Many commentators doubted that it could be done, but within three years oil producers had managed to persuade a divided Congress to overturn the ban.

The policy contest started slowly. In October 2013, Harold Hamm, chief executive of Continental Resources, an oil producer and shale pioneer, convened oil executives and journalists in Washington DC to make the case for restrictions on oil exports to be lifted (Harder and Berthelsen, 2015). Then in December 2013, the industry was given hope by remarks from then Secretary of Energy, Ernest Moniz, that it might be time for the US Government to review its 40-year ban on oil exports (Commodities Now, 2013). However, it was not until early 2014 that oil and gas producers began to publicly make the case. On 7 January 2014, Jack Gerard used the launch of the State of American Energy report by the API to call for an end to the 'arbitrary' and 'unfair limits' on oil exports (Gerard, 2014). The API was supported by the COC the next day (US Chamber of Commerce, 2014), and by the end of the month the US Senate had held the first hearing in 25 years to explore the 'opportunities and challenges' of crude oil exports, with the support of Republican Senator Murkowski, a vocal champion of the oil industry (US Senate Committee on Energy and Natural Resources, 2014). Further Congressional inquiries into energy exports followed as dozens of oil and gas executives travelled to Washington DC to meet legislators and propose legislation to remove the ban (Congress.gov, 2014; Lipton and Krauss, 2015).

One of the first hurdles oil producers had to overcome was the public concern about the economic impacts (Interview 42). As Senator Murkowski put it, 'there are still an awful lot of people who remember the gasoline lines we had in 1973' (Harder, 2014). And polling indicated that around 70 per cent of Americans opposed oil exports if they raise gasoline prices (Gebrekidan, 2014; Brown *et al.*, 2014). In response, as one lobbyist proclaimed, 'oil producers engaged in a textbook public campaign' laying the groundwork in 2014 'before they began their push in 2015' (Interview 16). The groundwork entailed funding a series of studies that countered the public concerns. For example, the API, which had been driving the campaign, along with Exxon-Mobil, Conoco Phillips, Continental resources and others, funded a report from the Aspen Institute that concluded that removing export restrictions would create more than half a million jobs with widespread economic benefits. There were similar reports from think tanks, such as the American Enterprise Institute, which also received funding from the industry (Lipton and Krauss, 2015). Citing many of these reports, the API released a press statement titled: 'Every Major Study Agrees: Crude Oil Exports Would Put Downward Pressure on US Gasoline Prices' (API, 2015).

In proclaiming the economic benefits, producers once again leveraged their discursive power and framed the solution as free trade (Interview 8). The IPAA argued that 'allowing for a freer oil market will boost American job creation, grow our economy, and secure our energy future' (Bell, 2014). The principle of free trade helped to tie together many of the industry associations that had supported gas exports, including the major oil and gas associations, namely the API, ANGA, and the IPAA, as well as the wider business community, such as the COC and the NAM. As one industry association executive pointed out 'we are pro-free trade and we are working to make it happen … we are working together' (Interview 35).

However, further up the supply chain some oil refiners were opposed. Refineries have traditionally been part of an integrated oil supply chain in the US, with companies such as ExxonMobil, Chevron, and Shell all maintaining refining capacity. However, following industry restructuring in the 1990s and 2000s the majority of refining capacity became independent (Khan, 2013). For the independent refiners the shale oil boom 'breathed new life' into refineries, which had been shutting down excess capacity until 2011. In addition to increased production, lower oil and gas prices also increased the international competitiveness of these operations (Warmann, 2015; Khan, 2013). As a result, according to oil refiners, an increase in the domestic oil price from 'lifting export restrictions and sending our crude overseas' would mean that US 'refineries would lie dormant once again' (Warmann, 2015).

Valero Energy was one of the first refiners to oppose lifting the ban, with its head William Klesse arguing that unfettered exports of oil and gas will raise costs (Klesse, 2013). However, the principal industry associations for refiners, the AFPM, was compromised because some of its members, such as ExxonMobil supported exports, while independent refiners like Valero Energy did not (Interviews 4 and 16). As a result, in March 2014 in the midst of a series of Congressional hearings on energy, four independent refiners – Philadelphia Energy Solutions, Alon USA Energy, PBF Energy, and Monroe Energy – launched the CRUDE Coalition (Consumers and Refiners United for Domestic Energy) to oppose lifting the ban (Mundy, 2014b). Challenging the free trade argument, the CRUDE Coalition argued that the 'world crude markets are not free markets' because they are manipulated by OPEC. Hence, lifting the export ban would result in oil being taken out of a competitive domestic market and put into less competitive global one 'controlled by the likes of Iran, Russia, Libya and other unfriendly regimes' (Warmann, 2015).

The opposition from refiners was countered by the oil industry. In September 2014, 14 oil and gas producers established Producers for American Crude Oil Exports (PACE), another ad hoc coalition whose stated purpose was to overturn the ban (Rowell, 2014). Comprising primarily independents, such as Andarko Petroleum, it provided a platform for these producers to add their voice to the campaign. It was complemented by their traditional industry association, the IPAA, which along with API, rounded on the refiners in the media and increased their lobbying efforts (IPAA, 2014). In March 2015, 16 CEOs from the industry and PACE met with the White House senior policy advisor, Brian Deese, to lobby the Obama Administration, which did not support lifting the ban (TOGY, 2015).

In the colourful words of one lobbyist representing refiners the campaign was:

A David and Goliath battle. We are fighting a war of attrition. We take a few shots and retreat, take a few shots and retreat. It is like the Minutemen versus the redcoats. The API has spent untold amounts of money mounting a campaign. It has been their main objective for a while.

(Interview 16)

For the independent oil producers that produce so-called sweet crude oil, which is low in sulphur, there is an especially strong commercial incentive to lift the ban. US refineries are designed to handle heavy crude oil, not sweet crude, and there is limited refining capacity to process the sweet crude. As a result, there is an oversupply of sweet crude on the domestic market, which means producers are forced to sell their product at a discounted price. While this suits refiners, it clearly does not suit producers, as one respondent put it, 'no good deed goes unpunished' (Interview 15). Although the CRUDE Coalition fiercely disputes many of these claims, it has been one of the principal arguments of producers like ConocoPhillips for lifting the ban (Lance, 2015).[7]

By 2015 the price spread between the domestic and international oil price had largely been erased as the global oil price tumbled. However, this did not stop oil producers continuing their campaign, which was boosted in July by a 12–10 vote on the Senate Energy and Natural Resources Committee in favour of lifting the ban, though the split was along party lines, with Democrat Senators opposing (Timothy, 2015). In September, the API launched a television advertising campaign in 12 states and the District of Colombia, focussed directly on exports (Dlouhy, 2015a). It also launched additional websites and funded networks of thousands of employees, such as Energy Citizen and Energy Nation, in support of lifting the restrictions. This was complemented by similar campaigns from other companies, including ConocoPhillips (Mikulka, 2015).

By October 2015, the geopolitical arguments that oil and gas producers had been making to support exports received an unexpected fillip. Firms such as Conoco Phillips had long argued that US oil would enable 'overseas allies to diversify their energy supplies, thereby strengthening US commercial and geopolitical influence' (Lance, 2015). On 18 October, President Obama announced that the US was preparing to lift sanctions against Iran following its agreement to curtail its nuclear program (*Guardian*, 2015). In other words, it would be permitted to sell its oil on the international market. This gave producers a powerful line of argument, which according to respondents resonated with many members of Congress: 'how can you lift sanctions on Iran but not on our own oil producers' (Interview 16). To further this line of argument, the industry also enlisted government leaders from the Czech Republic, Japan, and South Korea to 'communicate their support through diplomatic channels that they favour lifting the ban' (Lipton and Krauss, 2015).

After several false starts it appeared that by late 2015, oil producers were close to securing Congressional support to end the restrictions.[8]

However, the political environment was such that industry lobbyists worried about how they could get anything through Congress, especially given the significant opposition from Democrats and the White House. As one explained:

> [President] Obama is not going to change the regulations on oil exports on his own so we have to attach it to something.... We had been thinking the transport bill, but you need something that the Republican Party is going to support and something that will bring Democrats on board. But the transport bill was a no-go because it had 40 to 50 of the Tea Party against it.
>
> (Interview 35)

In the end, that something turned out to be the Omnibus Appropriations bill. On 18 December 2015, President Obama signed into law the Consolidated Appropriations Act of 2016, which included provisions that ended the 40-year restrictions on the export of crude oil. In return, Democrats secured an extension of production and investment tax credits for wind and solar energy, which I will turn to in Chapter 5 (Harder and Berthelsen, 2015). On 31 December 2015, the first shipment of crude oil was shipped from Texas to Italy (Carroll and Tobben, 2016).

Why and how did business actors shape the contest over oil exports?

Preferences

The contest over oil exports mirrored that over gas exports. Many of the same corporations formed a preference to support oil exports because it was in their commercial interests. For example, Shell, Exxon Mobil, and Chevron all supported lifting the 'unfair' and 'arbitrary limits', in the words of the CEO of the API, Jack Gerard, so that they could access international markets where oil was trading at a much higher price. In short, it would increase their profits. Further, given the uniformity in preferences across the oil industry there were no signs of intra-industry conflict, and the principal industry associations, such as the API, ANGA, and the IPAA all supported exports. Of course, in following their commercial interests firms invariably became entangled in inter-industry conflicts because the impact of lifting exports restriction was different on different industries. For example, in the refining industry, the principal independent refiners, such as Valero Energy, and members of the CRUDE coalition, opposed oil exports because any increase in oil prices would potentially increase their costs of production, reducing their international competitiveness and in turn their profits (Klesse, 2013).

However, there were exceptions in the refining industry, notably Phillips 66 and Marathon Petroleum, two of the largest refiners. Given that both these corporations' revenues derive primarily from refining it could be expected that they would also oppose the push to export oil. Instead, they appear to have supported oil exports or hedged their position (Volcovici and Gebrekidan, 2014; Meyers, 2014). The most likely explanation is the unique history of both firms (Levy and Kolk, 2002). ConocoPhillips and Marathon Oil, two large oil producers, which supported oil exports, had until recently maintained a refining capacity as vertically integrated oil companies. However, within a year of each other they split their refining operations, creating two new entities. Marathon Petroleum was established first in 2011 as a spin off from Marathon Oil, and Phillips 66 was established in 2012 as a separate refining entity from ConocoPhillips. The close historical ties these refiners had to oil producers likely influenced their decision not to oppose exports.[9]

Strategies

In the contest over oil exports, business actors in the oil and gas industries continued to rely on similar strategies to those they had employed in the contest over gas exports – see Table 3.2. Again, the effectiveness of these strategies was a function of their resources and political legitimacy. First, firms mobilised coalitions. The API led the campaign to permit oil exports and it tied together industry associations across the oil and gas sector, including the IPPA and ANGA, as well wider business coalitions, notably the COC and the NAM, which brought their considerable resources and networks to the campaign for oil exports. In addition, oil producers continued to leverage transnational networks to enlist state actors abroad, including government officials from the Czech Republic, Japan, and South Korea, who used their diplomatic channels to pressure Congress.

However, the campaign met resistance, especially among independent refiners, who in turn mobilised coalitions. Independent refiners established the CRUDE Coalition to coordinate their activities in opposition to exports. This ad hoc coalition was created as a temporary coalition to contest oil exports given that these firms could not rely on the backing of their principal industry association, the AFPM, which had come out in support of exports. Producers responded with the creation of a similar ad hoc coalition called PACE to focus solely on promoting oil exports. While some respondents dismissed these groups as 'unsophisticated organisations' (Interview 10), in part this was the nature of their appeal. Because they were not bound by the constraints of an industry association, they could focus all their energy on the specific contest that they had been established to address.

> Trade associations are tied down and they can't put enough lead on the target because of the diversity of membership. So we are creating these informal ad hoc coalitions. It's not new but it is happening more now.
>
> (Interview 16)

Table 3.2 Summary of the policy contest over oil exports

	Oil producers	Oil refiners
General industry preferences	Support	Oppose
Key coalitions mobilised in the energy sector	API, ANGA, IPAA, AFPM, PACE	CRUDE Coalition
Key business coalitions mobilised	COC and NAM	–
Other important actors in the networks	Foreign governments, e.g. Czech Republic	–
Framing strategy	Framed around free trade	Counter-framed that oil markets not free
Lobbying	Oil producers collectively spent hundreds of millions (e.g. ExxonMobil $25 million, Chevron $15.5 million)	Refiners collectively spent tens of millions (e.g. Valero Energy, $3 million)
Outcome	Restrictions on oil exports overturned	

What is striking about these coalitions is how nimble they are. Whereas traditionally scholarship has focussed on the role of large established industry associations, which wade into policy contests like elephants using their size and weight to ride over their opponents, here business actors are creating issue-specific coalitions, which suddenly appear to attack a particular contest before disappearing just as quickly. For well-resourced business actors, like ConocoPhillips, the information age and communication technologies enable them to quickly establish coalitions, such as PACE, which can then be disbanded when the contest is over.

A second strategy used to shape the contest was framing. Led by the API, oil producers framed the contest around jobs and growth and free trade. As the CEO of ConocoPhillips argued: 'it is time to let American oil trade freely on the global market' (Lance, 2015). As a strategic frame free trade was effective because it had normative appeal among other business coalitions. For example, one of the reasons that the AFPM, which represented independent refiners as well as producers, was reluctant to oppose crude oil exports was because of the free trade argument. As one respondent pointed out, usually the AFPM is 'in lockstep with refiners, but on oil exports there is a little bit of space' because 'they are a free market orientated organisation' (Interview 4). Similarly, the Business Roundtable, which is an especially influential business coalition, largely remained on the side-lines of the contest even though some of its members opposed exports because of its historical support for free trade (Interview 11).

Further, producers used the frame of free trade to link issues by drawing attention to the geopolitical advantages that would accrue from providing oil and gas to US allies in Europe and Asia, who were reliant on Russian oil. For instance, exploiting the US decision to lift sanctions on Iran, lobbyists argued: 'how can you lift sanctions on Iran but not on our own oil producers' (Interview 16). This line of argument also dovetailed with the pressure coming from foreign governments, which as noted, the industry had enlisted to support the campaign. In both cases these frames were contested, but more often than not they were contested on the terms set by oil producers. For example, the CRUDE Coalition, which led the refiners' opposition, attempted to claim that oil could never be traded freely in a global market controlled by OPEC (Interview 16).

Much like the contest over gas exports, oil and gas producers mobilised their overwhelming financial resources to fund their lobbying campaign. Between 2014 and 2015, for example, ExxonMobil spent around $25 million in lobbying and Chevron $15.5 million. Refiners could not match these resources. Valero Energy, one of the largest refiners, spent just under $3 million (CRP, 2015c). As the challengers to the status quo, producers relied heavily on outside lobbying to increase the salience of the issue and create incentives for policymakers to put it on the agenda (Layzer, 2007). The API coordinated this campaign via advertising, media, and grassroots mobilisation. For example, it led the 12 state advertising blitz and it launched the Energy Citizen rallies that took place in a number of states.

Of course, firms on both sides of the contests continued to conduct inside lobbying by targeting Congress. For example, Harold Hamm, the CEO of Continental Resources, made more than 30 trips to the capital and held more than 240 meetings with congressional members and their staff (Harder and Berthelsen, 2015). In doing so they also relied on their personal networks. For example, Republican Senator Lisa Murkowski, who chairs the Senate Energy and Natural Resources Committee, and has helped to get supporting legislation out of the committee, has worked closely with the oil industry in Alaska, her home state, and her former staffers are now employed in the industry (Interview 42).[10]

Finally, what impact did business actors have on the policy contest? The short answer is a significant influence. Indeed, producers seemingly had a greater influence on the contest over oil exports than they did over gas exports. Prior to the campaign from oil producers there was strong opposition to overturning a ban that had endured for 40 years with bipartisan support. Yet by the end of 2015 it was gone. It is unimaginable that this would have happened without the oil industry campaign, which was especially effective because of the significant financial resources the industry had at its disposal. It is not clear that other industries would have had the same success, and refiners appear to have had little impact, other than slowing down efforts to remove the ban.

However, as discussed above, the capacity of business actors to shape the outcome was once again primarily conditioned by the role of policymakers. The White House and most Democrats in Congress opposed lifting the ban, a position that derived from their beliefs and political incentives. For example, many Democrats espoused a belief that the various proposals on the floor of Congress too heavily supported fossil fuels (Timothy, 2015). And this belief was reinforced by the political incentives to oppose, with polling indicating that around 70 per cent of Americans opposed oil exports if they raise gasoline prices (Gebrekidan, 2014; Brown *et al.*, 2014). Of course not all representatives have the same political incentives, and this opened up opportunities for business actors in support of exports. For instance oil producers instead targeted representatives from oil-producing states, who had close ties to the industry to advance the campaign in Congress including Senator Murkowski, Republican of Alaska and Senator Heitkamp, Democrat of North Dakota. Others, such as Representative Barton, a Republican from Texas, home to many large oil corporations, supported the producers' campaign in the House. And there were big financial rewards. One report calculated that supporters of lifting the ban have on average received 155 per cent more money from associates of the oil industry than those opposed (Lipton and Krauss, 2015). In short, the strategies employed by business actors and their capacity to shape the contest were mediated by the political incentives of policymakers.

Conclusion

The shale revolution and subsequent surge in oil and gas production triggered two critical policy contests over the export of natural gas and crude oil. Business actors in the oil and gas industries led the campaign to ease restrictions on the exports of these commodities. In doing so, they actively worked to influence the policy process and shape the ultimate outcome in cooperation with, and in conflict with, firms across the US energy sector and the wider business community. As a result, these two contemporary policy contests provide an excellent window through which to consider how and why actors in these industries behave, something that has been largely missing in energy and environmental politics scholarship, but is critical for scholars and policy-makers considering future attempts to regulate these industries.

Table 3.1 and 3.2 summarised business preferences and strategies in both policy contests. In general, oil and gas producers supported easing restriction on the exports of these commodities because it was in their commercial interests. Exports would provide access to international markets and higher prices for their products. This led to inter-industry divisions as some business actors in the petrochemical industry and refining industry opposed gas exports and oil exports respectively, because in their view it was contrary to their interests. As discussed, to the extent that there were outliers, in other words, actors that developed a preference opposed to the rest of the industry, this appears to have reflected the unique history of some firms in the refining industry.

In both contests, these actors mobilised coalitions and leveraged networks to shape the outcome. Oil and gas producers mobilised key industry associations, notably the API and ANGA. These coalitions acted as the command centres of the campaigns as they worked to pool resources and coordinate activities. Some business actors also mobilised administration hoc coalitions that were established with the sole aim of shaping the contest over gas exports, such as America's Energy Advantage, or the contest over oil exports, such as the PACE coalition and the CRUDE coalition. The contests revealed that the formation of these ad hoc coalitions were particularly appealing to business actors in this sector because of their flexibility and focus. In addition, firms in both contests enlisted other business coalitions to the campaign, including the COC and the NAM, thereby building the resources and legitimacy of their campaigns. And, they leveraged transnational networks to enlist state actors abroad, including government officials to support exports of oil and gas.

In putting oil and gas exports on the agenda, producers defined the problem as restrictions on the exports of these commodities and framed the solution as free trade. As a normative frame free trade was effective because it had appeal among other business coalitions and policymakers in Congress. Framing was supported by the strategic dissemination of information via industry funded economic studies, which pointed to the economic benefits of freely trading oil and gas. The frame of free trade was also used to link issues by drawing attention to the geopolitical advantages of exporting oil and gas to US allies.

Further, drawing on their enormous financial resources, the oil and gas industry as a whole spent more than $140 million per year on lobbying during the years of the policy contests. The resources of oil and gas producers outweighed those in the petrochemical industry and refining industry. As one representative of the refining industry claimed, it was a 'David and Goliath battle', but these industries too spent millions seeking to shape the outcome (Interview 16). In both contests, firms engaged in inside and outside lobbying. The outside lobbying campaign was particularly evident in the contest over oil exports, as the API and its supporters worked to increase the salience of the issue via advertising, media, and grassroots mobilisation. Though in both contests inside lobbying was also a major focus, as firms targeted key policymakers in the Obama administration and in Congress.

Finally, the evidence indicates that business actors did influence the outcome of the policy contests. It seems unlikely that the decision to ease the restrictions on natural gas exports and crude oil exports would have occurred in the absence of a concerted campaign from oil and gas producers. This is especially so in the case of oil exports, given the historic opposition to oil exports. And there is good evidence to suggest that the resistance from some sections of the business community, such as petrochemical manufacturers in the contest over gas exports, was successful in limiting the speed with which producers were able to achieve regulatory change.

Nevertheless, the influence of business actors was mediated by the role of policymakers who, driven by their beliefs and political incentives, affected the capacity of firms to shape outcomes. For example, Democratic opposition to gas and oil exports, in part, reflected their concern about the fossil fuel industries contribution to climate change, but it also reflected concerns about the domestic economic impact of unfettered exports, which for actors opposed to exports, such as Dow Chemical, was an opportunity they could exploit via their lobbying campaigns. However, policymakers were also heavily influenced by their state-based political incentives, which is why oil producers, for instance, worked to enlist representatives of the House and the Senate who came from oil-producing states and were more likely to support the campaign of the oil industry. In short, the role of policymakers in each of the contests created and restricted opportunities for industry to determine the ultimate outcome.

In summary, these policy contests reveal much about the preferences and strategies of US oil and gas industries, including the divisions between industries, which could potentially be exploited for political gain – a topic I will return to in the final chapter. While there were no doubt other factors at play in determining the outcomes of both policy contests, which are outside the scope of this analysis, it is clear that business actors played a critical role.

Notes

1 Gerard (2015).
2 Following the policy contests in 2016, the API took over ANGA to form a single industry association (Cocklin, 2015).
3 Interview 20.
4 This was confirmed by other industry associations that claimed that because they 'had previously supported free trade 95 per cent of our memberships supported it' (Interviews 7 and 35).
5 Interview 16.
6 Under the powers of the *Energy Policy and Conservation Act of 1975*, administered by the Department of Commerce, oil producers in the US are restricted from exporting crude oil with a few exceptions, such as oil to Canada or oil produced in Alaska.
7 For further background on the policy issues (see for example, GAO, 2014).
8 In October 2015, the House passed a bill to allow exports, and the Senate Energy and Natural Resources Committee, chaired by Senator Murkowski, had twice approved legislation permitting oil exports (Reuters, 2015, Dlouhy, 2015b).
9 It should also be noted that large oil and gas producers, such as ExxonMobil and Shell, also had refining capacity as vertically integrated companies, but they remained predominantly upstream companies given the large revenues they earned from oil production, which meant they had much to gain from exports.
10 Senator Murkowski has also been a strong opponent of President Obama's proposal to protect Alaska's Arctic National Wildlife Refuge, which would prevent oil companies from drilling in the area (NPR, 2015).

4 The war on coal

Policy contests in the coal and utility industries

Introduction[1]

'It is a democratic war on coal'[2]

Under President Obama the coal industry and the utility industry had been warning of the casualties of a 'war on coal'. According to the industry, the so-called war reflected a series of regulatory initiatives that President Obama supported to target coal, the largest contributor to greenhouse gas emissions in the US. However, the US coal industry is also fighting on a second front: the industry is in structural decline. Coal production is declining, the number of producing mines is declining, productive capacity is declining and the number of employees at US coal mines is declining (EIA, 2018a). The structural decline of coal is also being felt in the utility industry. Over the last two decades, coal's share of electricity generation has fallen from more than half in 1990 to around a third today (EIA, 2017b: 60). And, record numbers of coal-fired power plants are now retiring, with almost 100 retiring in 2015 alone, in part due to Obama-initiated regulations, such as the Mercury and Air Toxic Standards (MATS) (EIA, 2016b). Most projections expect the retirements to continue in the coming decades (EIA, 2018a).

One of the key reasons for the structural decline of coal is the shale revolution. As discussed in the last chapter, the shale revolution has led to historically low gas prices, which means that in many cases gas is now more competitive than coal. And, because electric utilities absorb almost 90 per cent of US coal production, any move by utilities to substitute coal with gas has a deleterious impact on the industry (EIA, 2016f). Yet this is precisely what is happening. In 2012, the natural gas price fell below $2 Btu, which resulted in an unprecedented degree of switching in the utility industry. In fact, output from gas-fired plants surpassed that from coal plants in 2017, accounting for 32 per cent of primary electricity generation (EIA, 2018c).

While the future of coal may not be as positive as oil and gas, the US still holds the world's largest coal reserves, and, after China, it is the largest coal consumer in the world. Accordingly, if the world is to achieve an energy revolution, the regulation of coal in the US will be crucial. Indeed resistance from these industries could delay and even derail government attempts to address climate change. In order to consider how and why business actors in the coal and utility industries are shaping the rules that govern coal, this chapter will focus on two standout policy contests during the Obama administration. The first was the attempt by the new president in 2009 to establish a nationwide emissions trading system, otherwise known as the Waxman–Markey bill. The second was the Clean Power Plan designed to restrict emissions from power plants under the authority of the Clean Air Act. As the rhetoric of war suggests, business actors have fiercely contested these policies, especially in the coal and utility industries, and their active engagement provides a perfect opportunity to examine their preferences and the strategies they employ to influence policy outcomes.

The next section provides an overview of the US coal and utility industries, including the business actors. Following the structure of the last chapter, the sections thereafter examine the preferences and strategies of these actors in the contests over the Waxman–Markey bill and the Clean Power Plan.

Overview of the US coal and utility industries

The structural decline of the US coal industry shows no signs of ending and the impacts are being felt among coal producers and electric utilities. First to the coal industry. As discussed, the US has the largest recoverable reserves of coal on the planet and it is the second largest coal producer, accounting for 12 per cent of world coal production (IEA, 2016b: 230). However, the fortunes of King Coal are deteriorating. In 2015, coal production fell 10 per cent, the lowest annual production level since 1986. In the Appalachian region alone, coal production fell a staggering 18.6 per cent between 2015 and 2016 (EIA, 2017a: vii). In its reference case, the EIA projects that US coal production will continue to decline to 2022 due to the retirement of coal-fired power plants and competitive prices from natural gas and renewables (EIA, 2018d).

In the US electric utility industry, coal has been the number one source of electricity generation for more than 60 years. However, this, too, is changing. In 2016, natural gas overtook coal as the primary source of electricity generation for the first time. In 2017, natural gas comprised 32 per cent of electricity generation, coal 30 per cent, nuclear 20 per cent and renewables, including hydropower, 17 per cent (EIA, 2018e). Coal's declining share of electricity generation, which peaked at almost 60 per cent in the mid-1980s, is based, especially in recent years, on strong competition from cheap natural gas and a series of policy initiatives to reduce air pollution in the utility industry. As Figure 4.1 shows, coal's share of electricity generation will decline to 2040 because of unfavourable economic

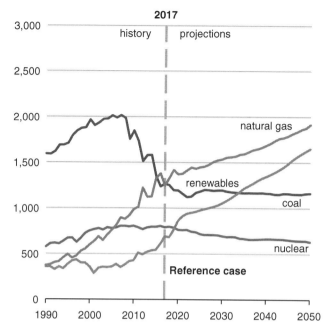

Figure 4.1 US electricity generation from selected fuels.

Source: Energy Information Administration (EIA, 2018a).

conditions compared with natural gas and renewables, though higher natural gas prices may see it increase in the short-term. Renewable generation is projected to overtake coal's share of electricity generation in around a decade (EIA, 2018a).

Declining coal production has had a significant impact on coal producers and electric utilities. For coal producers one statistic stands out: by 2015, more than 50 firms representing 50 per cent of US coal production had filed for bankruptcy protection (IEA, 2016b: 204). This includes three of the largest coal corporations – Peabody Energy, Alpha Natural Resources, and Arch Coal (Kary *et al.*, 2016). The US is not alone. In China, around 80 per cent of coal producers incurred losses in 2015. As the IEA points out, the profitability of the coal industry in some of the largest coal producing nations, such as the US and China, remains in question and excess global capacity means that exports markets offer little relief, including for US producers facing falling domestic demand (IEA, 2016b: 204).

For the utility industry the decline in coal is evident in a wave of coal-fired power plant retirements. In 2015, almost 15 GW of coal-fired capacity was retired and a further 45 GW is expected to be retired by the end of the decade. To put this in perspective, in 2015 coal retirements represented

80 per cent of total retirements in the electricity sector. Most of the coal plants that were retired were built between 1950 and 1970, with an average age of 54 years (EIA, 2016b). This has largely been driven by the MATS regulations, which came into effect in April 2015 to reduce pollution from the sector. Since 2015, natural gas has traded at $3.50 per MBtu on average, and is expected to remain competitive with coal-powered electricity generation to 2020 (EIA, 2018c). As a result, average electricity prices for industry in the US are lower than in China and Japan, and well below European nations, providing an important boost to US industrial competitiveness (IEA, 2017c).

Despite the structural decline in coal, current levels of coal production and coal electricity generation must fall further if the US is to contribute to limiting global temperatures to 2°C. Under its 450 (or 2°C) scenario, the IEA projects that coal-fired generation, which comprised 41 per cent of world electricity generation in 2014, will plummet to 7 per cent by 2040. In addition, under this scenario 70 per cent of coal-fired generation is equipped with CCS technology, mostly in China and the US – 'a technology yet to be developed and deployed at scale' (IEA, 2016b: 249). As a result, emissions from the electricity sector are projected to decline significantly at an annual average rate of 5 per cent. By 2040, global emissions from the electricity sector will be just over a quarter of 2014 emissions (IEA, 2016b: 250–251). While the projected decline in coal is more dramatic than that of oil and gas under this scenario, the risk of stranded assets in the coal sector is present but more limited because coal production is less capital intensive than oil production. Instead the risk of stranded assets, according to the IEA, 'is concentrated further down the value chain in coal-fired power stations' (IEA, 2016b: 154).

So who are the business actors? In the period of the policy contests in the US Peabody Energy, Arch Coal, and Alpha Natural Resources, made up the top three coal companies based on production, and they are followed by well-known companies including Cloud Peak Energy, Rio Tinto, Murray Energy, and Westmoreland Coal, among others. Peabody is the largest coal company and produced around 183 million short tonnes of coal in 2013 (EIA, 2015a: 16). Most of its coal assets are steam coal, which means that like many large US coal companies, it is acutely exposed to changes in the electric power sector given that steam coal is used to provide electricity (Witter, 2015c: 26). In the US, the majority of coal mined is located in West Virginia, Kentucky, and Wyoming. The American Coal Council (ACC) and the National Mining Association (NMA) have been the principal advocates for coal producers. Because of the diversity of membership, members of these associations are also often members of smaller more focussed groups as well, such as the American Coalition for Clean Coal Electricity (ACCCE), which was formed in 2008 to promote CCS (Interview 10).

Alongside the coal producers are the electric utilities. The utilities industry in the US is dominated by investor-owned utilities, which comprise about 70 per cent of the sector. Although the generation portfolios of these utilities vary, many use coal, given that coal remains the largest source of energy generation. In the period of the policy contests, Duke Energy is the largest investor-owned utility based on market capitalisation, followed by NextERA Energy, Dominion Resources, Southern Company, Exelon, American Electric Power, Sempra Energy, and PG&E, among many more. The principal industry association for the investor-owned utilities is the Edison Electric Institute (EEI), which has over 250 members operating across the US (EEI, 2014). There are also other smaller more focussed associations, such as the Nuclear Energy Institute (NEI), which represents utilities with nuclear power, or for example, the Electric Power Supply Association (EPSA), which represents utilities in the wholesale market (Interview 23).

Historically, the utility industry has been dominated by regulated utilities, which were designated as natural monopolies and whose rates were determined by public utility commissions (PUCs). However, in the 1990s the industry was partly deregulated with the *Energy Policy Act of 1992*, which allowed competition in the wholesale market. The effect was to transform the industry as firms cut costs, merged, and looked to purchase assets overseas (Hirsh, 1999). Though the wholesale market was deregulated, the retail market, where utilities distribute power to retail customers via the electricity grid, remains regulated. State governments grant investor-owned utilities a monopoly in a specified area and require utilities to provide electricity to all customers in that area at a reasonable price. PUCs set retail prices based on the cost of generation, infrastructure, and other expenses, and utilities are granted a specific rate of return on their investment in transmission, distribution, and generation infrastructure. While the retail market has provided some certainty to the incumbent utilities in the face of fierce competition in the wholesale market, as I will discuss in the next chapter, there is also growing competition in the retail market from renewable power.

The policy contest over emissions trading: 'it was clear that there was going to be legislation and legislation could succeed'[3]

The inauguration of President Obama in January 2009 came with high hopes of federal action on climate change. The President had campaigned strongly on the issue and polls showed that the majority of Americans supported action on climate change (Leiserowitz *et al.*, 2008). Within six months of taking office, the House of Representatives had passed the first ever federal legislation to mitigate climate change. The American Clean Energy and Security Act of 2009, the so-called Waxman–Markey bill. Named after its co-sponsors, it aimed to put in place a nationwide emissions trading scheme, or in American vernacular, 'a cap and trade' scheme

(Broder, 2009b). However, the legislation never made it through the Senate and 12 months later, after a bitter fight, the high hopes had been dashed and cap and trade was dead.

Business actors had been central to the fight, none more so than the coal industry and the utilities industry, which battled over the impact of the legislation. The Waxman–Markey bill comprised a series of measures, such as renewable energy and energy efficiency standards, designed to reduce greenhouse gas emissions by 17 per cent below 2005 levels by 2020. The centrepiece of the legislation was an emissions trading scheme, which set a cap on greenhouse gas emissions from covered sectors, such as the electricity sector and allowed these firms to trade the emissions permits. Electric utilities would be allocated the vast majority of their permits for free, with approximately 20 per cent of the schemes permits auctioned (Pew Center on Global Climate Change, 2009). While estimates varied, the EPA projected that the price of permits would be relatively modest at around $15 per tonne in 2015 (EPA, 2009b). The bill also provided financial incentives and federal funding for carbon, capture, and storage (CCS) technology (C2ES, 2009). Despite the battle over the bill, emissions trading was not new in the US. The EPA had experimented with emissions trading programs as early as 1974 and by the 1990s emissions trading programs were widely viewed as the best instrument for dealing with a range of environmental problems.[4]

The contest over federal cap and trade legislation had begun before President Obama came to office. On 15 January, five days before his inauguration, the US Climate Action Partnership (USCAP), a coalition of business actors and environmental NGOs, released their *Blueprint for Legislative Action*, which called for the 'prompt enactment of national legislation in the United States to slow, stop, and reverse the growth of greenhouse gas emissions over the shortest time reasonably achievable' (WRI, 2009: 1). USCAP's support for emissions trading had been two years in the making. Led by environmental NGOs and established in January 2007, the coalition brought together Fortune 500 companies, including major players in the energy sector, such as BP, ConocoPhillips, Duke Energy, and Shell.[5] Jim Rogers, CEO of Duke Energy, which at the time operated around 20 coal-fired power plants, was a crucial player (Bartosiewicz and Miley, 2013: 22–26). For many of these companies, such as PG&E, another major utility, the motivation for joining was to shape regulation because, according to one of the founders of USCAP, there was not a single USCAP CEO who did not consider future climate change regulation inevitable (Vormedal, 2011: 11; Bartosiewicz and Miley, 2013: 26).

USCAP was not the only business group with coal and utility industry representation. The ACC and the NMA, two of the preeminent coal associations opposed the bill (Broder, 2009a). In addition, the ACCCE, which promoted clean coal, actively opposed the Waxman–Markey bill. ACCCE comprised many of the most powerful coal corporations and electric utilities in the US including Peabody Energy, Southern Company, and American Electric Power (Interview 10). At the start of 2009 Duke Energy was a member of both, but it soon left.

The key point of the blueprint released by USCAP was the allocation of carbon permits. The issue was whether carbon permits would be given to utilities for free, or whether they would be auctioned, and utilities would have to buy them in the market. For Duke Energy and the utility industry, which produced around 40 per cent of US carbon dioxide emissions, free allocation of permits was a must. As CEO Jim Rogers made clear in the USCAP negotiations, Duke Energy was not going to sign on to the blueprint unless it gave 40 per cent of the allowances to the power sector for free. Oil and gas corporations, led by ConocoPhillips, resisted the utility industry position on free permits, but with Duke Energy threatening to walk out if this was not included, they decided to remain at the table and support the blueprint rather than walk away (Pooley, 2008: 318–319).

However, in February 2009 President Obama released his first budget, which called on Congress to enact an emissions trading scheme with 100 per cent of permits to be auctioned, raising almost $80 billion in revenue by 2012 (Chapman and Dodge, 2009). In response, the utility industry went into action. Led by Duke Energy they launched a press offensive and began lobbying on the hill, meeting with House and Senate representatives, including with representative Henry Waxman, the newly installed chairman of the Energy and Commerce Chairman and one of the architects of the bill (Pooley, 2008: 337–339). By the time the draft legislation was released at the end of March, the finer details, including the allocation of permits, were left open.

With EPA analysis showing that the Waxman–Markey bill would largely stop the construction of new coal power plants, it was with the coal industry and utility industry that legislation faced its toughest negotiations (EPA, 2009b). As part of USCAP, Duke Energy, and the EEI coordinated the utilities position and directly lobbied Congressman Waxman's staff. After lengthy negotiations, the bill introduced into the House provided the utility sector with 35 per cent of allowances for free, enough to cover 90 per cent of its emission (Pooley, 2008: Part 9). The bill also included $1 billion in funding to support CCS, alongside a tray of other sweeteners. This proved enough for the EEI to support the bill, and Tom Kuhn, the association's executive director, actively lobbied in support of the bill (Mildenberger, 2013: 28–29).

Yet not all of the utility industry was on board and many utilities took different positions, which were closely correlated to their generation portfolios. For example, the EEI's ultimate support for the Waxman–Markey bill led to 'internal dissent' within the EEI and a vocal group of utilities calling themselves the Midwest Climate Coalition actively campaigned against the bill (Mildenberger, 2013: 29). Others, such as Southern Company, had strongly resisted the agreement struck among the EEI utilities, and it lobbied harder than anyone to dilute the bill (Pooley, 2008: 371–372). In fact, Southern Company hired 63 lobbyists to work on the bill. In comparison, American Electric Power, a similarly sized utility, hired nine lobbyists (Lavelle and Donald, 2009). Yet utilities with more to gain from emissions trading, due to their renewable portfolios, such as PG&E, campaigned alongside USCAP for the bill (Kim *et al.*, 2016: 23).

With the support of USCAP and EEI, the Waxman–Markey bill narrowly passed the House of Representatives on 21 May 2009, with 219 votes in favour to 212 against, only one vote more than the 218 votes required for a majority (Broder, 2009b). While President Obama, the Leader of the House Nancy Pelosi, along with Representatives Waxman and Markey championed the bill, 44 Democrats voted against it, mostly from coal producing and industrial states. Only eight Republicans voted in favour (Sheppard, 2009).

The bill had also split business coalitions, but as negotiations moved to the Senate the divisions widened. Within days, the membership of USCAP began to splinter. Caterpillar Inc., a global producer of mining and construction equipment, withdrew its support for the Waxman–Markey bill, so, too, did ConocoPhillips, which was aggrieved that the oil and gas sector received too few emission permits (Bartosiewicz and Miley, 2013: 55).

At the same time, ACCCE continued to campaign against emissions trading. Despite the sweeteners for coal, the problem for the ACCCE was that the scheme would commence before clean coal technology was commercially viable. Reflecting its financial resources, the ACCCE had spent almost $15 million in lobbying activities, funded by some of the most powerful coal companies in the US, to undermine federal climate legislation, but its campaign was now coming undone (Kim *et al.*, 2016: 2). The group had been caught in a public relations scandal following revelations it funded astroturfing activities, hiring a public relations firm to send fake letters to congressional members opposing the Waxman–Markey bill, supposedly from groups such as the National Association for the Advancement of Colored People (GreenBiz, 2009). By September 2009, Duke Energy, which had been a member of the ACCCE, left the group. It was followed by Alcoa and First Energy (Pooley, 2008: 410). Tension was also mounting in the COC, which had been outspoken against the bill. While Duke Energy remained a member, other utilities exited USCAP, including Exelon, PG&E, and PNM. Exelon's CEO, John Rowe, said that the utility would not renew its membership due to the Chamber's 'stridency against carbon legislation' (Krauss and Galbraith, 2009).

In the Senate, a series of attempts to introduce a comparable bill followed. The first, introduced by Democratic Senators Kerry and Boxer in September 2009, had been stripped of reference to cap-and-trade in the drafting process and was branded a 'pollution reduction and investment' program. Another attempt came in the form of a cap-and-dividend policy proposed by Democrat Senator Cantwell and Republican Senator Collins.

Finally, in May 2010, Democrat Senator Kerry and independent Senator Liebermann produced an updated version of the Kerry-Boxer bill, which many believed had the best prospect of succeeding. However, the oil and gas industry and sections of the coal and utility industries were determined to ensure that it did not (Pooley, 2008: 410). The API led the charge. It began spreading reports about the huge economic costs and job losses that emissions trading would wreak (Vormedal, 2011: 22). Then in August, Jack Gerard, president of the API, called on its members to send their employees to 'Energy Citizen' rallies, which were held in swing states around the country. At the first Energy Citizen rally in Houston 3,500 people participated (Pooley, 2008: 410). In the end, the draft senate bill established an emissions trading scheme for electric utilities with a 'fee' linked to the price of permits for oil companies (Vormedal, 2011: 23). In other words, oil and gas companies were removed from the emissions trading scheme. According to one insider:

> This was the grand bargain that we struck with the refiners. We would work with them to engineer this separate mechanism in exchange for the American Petroleum Institute being quiet. They would not run ads, they would not lobby members of Congress, and they would not refer to our bill as a carbon tax.
>
> (Lizza, 2010)

After years of negotiations, hundreds of millions spent in lobbying and a bitterly divided business community, in July 2010 the Democrat majority leader, Senator Harry Reid, announced that the US Senate would not consider legislation for an emissions trading scheme. As he put it, 'we know that we don't have the votes' (Hulse and Herszenhorn, 2010). With that, cap and trade was dead, and so, too, was USCAP, which after three years of negotiations disbanded.

Why and how did business actors shape the contest over emissions trading?

Preferences

In the contest over the Waxman–Markey bill coal producers and utilities formed a preference to support the bill, oppose it, or hedge their position, which to a large extent reflected their commercial interests. First, in the coal industry there was almost uniform opposition to the Waxman–Markey bill, with Rio Tinto the principal exception. Although the projected cost on coal was arguably modest, at around US$15 per tonne, according to EPA estimates, all coal producers stood to lose (EPA, 2009b). Further, the history and culture of the coal industry reinforced this position. The coal industry has a long history of opposing environmental regulations, especially climate initiatives, and many of their industry associations, such as the NMA and the ACC, derided the need for action on climate change and saw the Waxman–Markey bill as part of an ideological crusade, a 'war on coal' (Interviews 5 and 51). Given the uniformity in preferences, as expected there were no signs of intra-industry conflict, and the principal coal associations the ACC and the NMA campaigned against the legislation (Interviews 5 and 51).

The exception was Rio Tinto. There are two potential explanations; first, Rio Tinto was less exposed to coal than other coal corporations. Coal contributed just 8 per cent to Rio Tinto's total global revenues compared to more than 90 per cent for most coal corporations (Rio Tinto, 2010: 5). Second, and arguably more importantly, were institutional factors. Research on business actors indicates that the home country of a corporation and the associated political, economic, and cultural context can shape business decisions (Falkner, 2008; Levy and Kolk, 2002). This appears to be the case with Rio Tinto, which is headquartered in the United Kingdom. Unlike its counterparts headquartered in the US, Rio Tinto was more familiar with emissions trading. It had been exposed to similar debates in Europe and Australia and had come to support a market-based mechanism in these jurisdictions prior to the contest over the Waxman–Markey bill (Rio Tinto, 2009).

In the utility industry commercial interests also drove business preferences, although they varied according to generation portfolios. In other words, preferences varied to the extent to which they relied on coal to produce electricity. On the surface the largest electric utilities appear to have supported the Waxman–Markey bill. However, a closer interrogation of the data indicates that, in reality, utilities can be categorised into three broad groups, which reflect their generation portfolios. First, utilities that maintained diverse generation portfolios, which included less than a third coal and often had growing renewable energy portfolios openly supported the regulations because of the relative gains they were likely to make, such as NextEra Energy and PG&E (Point Carbon 2009).

Second, utilities that generated more than a third of their electricity from coal, and often as much as two-thirds, such as the largest utility in the US Duke Energy, hedged their position. With strong pressure for regulatory action on climate change under President Obama, they worked to shape the legislation to minimise the compliance costs. This meant that as some points they opposed the Waxman–Markey bill and at others they supported it as they worked to shape the regulations. This reflected the view among many business actors that climate change regulation was inevitable. As the CEO of Duke Energy Jim Rogers put it, 'if you're not at the table, you're going to be on the menu' (Smith and Bustillo, 8 July 2012). American Electric Power was a similar case. It also relied on coal for close to two-thirds of its generation capacity. While it ultimately decided to support the Waxman–Markey bill, interviews with utilities suggested that this was only after they became convinced that legislation was inevitable (Interview 6). As the CEO of American Electric Power, Mike Morris, stated:

> American Electric Power supports the American Clean Energy and Security Act. No legislation is perfect – particularly one that seeks to overhaul the way our nation uses energy – but we believe this climate bill will work and it represents the best of the available options.
>
> (American Electric Power, 2009)

Third, some utilities that generated more than a third of their electricity from coal decided to oppose the Waxman–Markey bill outright rather than hedge their position. Southern Company was the principal utility to adopt this position with coal representing more than two-thirds of its generation capacity (Southern Company, 2016). The interview data suggests that it was especially close to the coal industry and strongly supported its position against the bill. In addition, unlike Duke Energy, for example, it was not a member of USCAP, which supported emissions trading (Interviews 46 and 51).

This variation in preferences did result in intra-industry conflict. While the leading industry association, the EEI, supported emissions trading, other EEI members broke ranks and campaigned against their industry colleagues. Southern Company, for instance, spent millions lobbying against the legislation and they were joined by the Midwest Climate Coalition, which also split (Mildenberger, 2013). These divisions spilled out across the business community. Some utilities, such as Exelon and PG&E, which backed the bill, abandoned their membership of broader industry associations, notably the NAM because of its position against emissions trading (Morford, 2009; Interviews 35 and 14).

Strategies

To achieve their preferences, business actors in these industries built coalitions and leveraged networks, which in turn were used to frame the contest and to lobby – see Table 4.1. As discussed in Chapter 2, the effectiveness of these strategies to a large extent was a function of the resources business actors had at their disposal and their political legitimacy. First, the most important coalition was USCAP, which led the campaign for emissions trading by tying together coalitions across the business community, such as EEI, the leading utility association. In the language of coalitions, USCAP was a key part of the 'advocacy coalition' that helped to build momentum for the bill (Sabatier, 1988; Knox-Hayes, 2012). USCAP was influential not only because it counted significant corporations from the oil, gas, and coal sectors among its membership, but because it was a classic example of a coalition of 'Baptists and bootleggers' in which

Table 4.1 Summary of the policy contest over emissions trading

	Coal producers	Electric utilities
General industry preferences	Oppose	Support/hedge/oppose
Key coalitions mobilised in the energy sector	ACCCE, ACC, NMA	USCAP, EEI
Key business coalitions mobilised	COC and NAM	COC and NAM
Other important actors in the networks	–	Environmental NGOs and unions
Framing strategy	Framed around economic costs and job losses	Framed around economic costs and job losses
Lobbying	The coal industry spent $34 million on lobbying between 2009 and 2010	The utility industry spent $336 million on lobbying between 2009 and 2010
Outcome	Emissions trading defeated	

environmental organisations and business actors campaigned for the same goal even if for different reasons (Desombre, 1995).

This network of business actors also met resistance from the coal industry, sections of the utility, and oil and gas industries, which mobilised coalitions. This included the ACCCE, which led the opposition campaign, the ACC and the NMA. These industry associations in turn leveraged the support of the COC and the NAM. Many of these actors, both individual firms and coalitions, also worked to enlist other actors that would support their cause and build legitimacy, which meant an ever-widening network. For example, American Electric Power one of the largest electric utilities, leveraged its historical ties to the union movement, whose members were employed in its power plants, to enhance the legitimacy of its position with Democrats in Congress (Interview 6). Similarly, oil and gas corporations enlisted their employees in so called 'Energy Citizen' rallies held in swing states around the country to enhance the public legitimacy of its campaign against emissions trading. The mobilisation of volunteers by state actors to achieve governance outcomes is not uncommon in other policy domains, for example police have enrolled citizens to achieve security objectives, but the evidence here suggests that business actors are now doing the same (Wood and Shearing, 2007: 17).

In attempting to exercise influence over the policy process, business actors across the coal and utility industries used their discursive power to frame the debate and set the agenda. This was most evident among business actors in the coal and utility industries opposed to the legislation, which defined the Waxman–Markey bill as the problem and framed the contest around economic costs and job losses. For example, coal producers consistently labelled emissions trading as an economic threat highlighting the job losses that would flow from its implementation (Peabody Energy, 2014). Mirroring the oil and gas industry, business coalitions relied on a series of economic studies to inform public debates and disseminate information to Democrats and Republicans in Congress. A widely cited study by the Heritage Foundation, a conservative think tank, estimated that the Waxman–Markey bill would slash real GDP by $9.4 trillion and increase unemployment by almost 2.5 million by 2035 (Beach *et al.*, 2009). While such estimates were vastly different to those of the EPA, they helped to strengthen the strategic frame.

Lobbying was also ubiquitous throughout the policy contest. To be sure, in 2009 the Waxman–Markey bill was the 9th most lobbied bill among all federal legislation (Kim *et al.*, 2016: 3) and coal producers and electric utilities, not to mention the rest of the business community, leveraged considerable financial resources. Between 2009 and 2010 coal producers spent $34 million on lobbying and utilities $336 million (CRP, 2016c, 2016e). Key business coalitions coordinated much of the lobbying activity including USCAP, which spent close to $3 million on lobbying over the course of the contest and the ACCCE, which spent close to $15 million (CRP, 2016b,

2009). Of course, consistent with previous research, most corporations and business coalitions had their own lobbying arms and had established political action committees (PACs) to mobilise resources (Vogel, 1989). These resources were used for both inside lobbying, such as directing donations to politicians, and outside lobbying, such as advertising campaigns and other activities, including astroturfing, which ended in scandal for the ACCCE.

Finally, what influence did business actors have on the policy contest? In examining business preferences and strategies it is evident that this is not an easy question to answer. For example, while the support of major utilities, such as Duke Energy, and their principal industry association, the EEI, was likely crucial to getting the bill across the line in the House, the vociferous opposition from coal producers and evolving opposition from other industries, including oil and gas, was equally damaging, and helped to kill any comparable bill in the Senate.

In addition, the influence of these actors on the policy contest was conditioned by the mobilisation of other non-state actors and the role of policymakers. First, in the contest over emissions trading the mobilisation of environmental NGOs via USCAP likely aided the capacity of electric utilities who were members of the coalition to shape the bill. This is because by working with environmental NGOs, business actors, including electric utilities, were not only able to increase the resources at their disposal, but also to enhance the legitimacy of their position given that they could demonstrate to policymakers the breadth of support for the USCAP position. For similar reasons, it is also likely that the mobilisation of these actors therefore limited the influence of coal producers, who could not rely on similar levels of support from other non-state actors outside the business community.

Second, the role of policymakers also shaped the opportunities for business actors to influence the contest. To be sure, it was President Obama's election, together with the Democratic takeover of the House, which led many business actors, including those in the utility industry, to conclude that climate legislation could succeed and hence it was necessary to engage. The White House and Democrats in Congress had espoused a belief in the need for climate action and there were political incentives to do so, with polling showing that the majority of Americans supported efforts to curb emissions, though critically, concern about climate change did not rank as a top priority among voters (Egan and Mullin, 2017: 355; Layzer, 2012). This is one of the reasons coalitions, such as USCAP, were active developing plans for an emissions trading scheme prior to the President's inauguration. For those outright opposed, such as coal producers, part of their lobbying effort focussed on representatives from coal-producing states. It comes as little surprise therefore that of the 44 Democrats in the House who voted against a bill championed by their own party, the majority came from coal-producing and industrial states (Sheppard, 2009).

Likewise, the oil and gas industry was able to target representatives from oil and gas producing states to ensure than any legislation had a limited impact on their operations (Lizza, 2010).

The policy contest over the Clean Power Plan: an 'existential crisis' for coal[6]

In the wake of the failure to legislate an emissions trading scheme, President Obama turned to regulation in his second term. In June 2013, he directed the EPA to establish carbon pollution standards for power plants, with the aim to reduce emissions from the utility industry by 30 per cent from 2005 levels by 2030 (The White House, 2013b). The announcement was the central part of the President's Climate Action Plan, which comprised a series of initiatives to ensure that the US would meet its commitment to reduce greenhouse gas emissions by 17 per cent below 2005 levels by 2020 (The White House, 2013a). The decision to turn to regulation after the failure of the Waxman–Markey bill was made possible by a succession of rulings by the Supreme Court, which confirmed the EPA's power to regulate greenhouse gases.[7]

The President instructed the EPA to establish carbon pollution standards under the Clean Air Act both for new and existing power plants. Under the Act (section 111b), the EPA was to establish a federal standard based on the best demonstrated technology for new or modified power plants, which in effect requires new power plants to operate with CCS technology. However, for existing power plants (section 111d), which are the largest source of carbon dioxide emissions, the EPA was to set emission guidelines with states to design individual implementation plans, which are consistent with the EPA's guidelines. In doing so, the EPA provided significant flexibility for states to reduce emissions from power plants using four 'building blocks', including increasing the efficiency of plants or relying on natural gas.[8] Where states failed to submit a satisfactory plan, the EPA could design and enforce a federal plan to reduce emissions from existing power plants in that state (EPA, 2015).

At a meeting at the offices of the Business Roundtable shortly after the President's announcement, Mike Duncan, CEO of ACCCE, argued that the plan will mean 'higher energy costs, less reliable electricity, lost jobs and a shattered economy' (Duncan, 2013). ACCCE was supported by many of the largest coal producers, such as Peabody Energy and Alpha Resources (Maher, 2013). It was also supported by electric utilities, such as Southern Company, which had fought emissions trading (Interview 46). Yet not all utilities were on board. For example, Duke Energy, which had been active in USCAP, said it was prepared to work with the EPA (Downey, 2013), as was Exelon, which favoured regulation because of its nuclear fleet (Interview 24).

Nevertheless, much of the coal industry, the utility industry and for that matter the broader business community mobilised to fight the regulations, not to mention the Republican Party, which now controlled the House following the mid-term elections in 2010, and was deeply antithetical to any regulatory action on climate change. In the same speech Mike Duncan set out the strategy.

> The fight before us will come in two stages, one inside the Beltway and one outside. The first round will be fought here in Washington, as public comments are gathered. The second will take place at the state level, as state governments develop plans to meet the proposed standards.
>
> (Duncan, 2013)

And the first round started right away. In the early EPA 'listening sessions', which gathered public comments, business lined up to critique the proposal. For example, at one session in November 2013 the COC argued that 'it is difficult to overstate the economic threat presented by these regulations, which together with other EPA rules could wipe out an entire industry' (Harbert, 2013). In the statements that followed, companies highlighted the potential costs, argued for more flexibility and delayed implementation. The aim of the industry, as one executive conceded was 'to build technical comments to undermine legislation' because 'everything we do is to build a record in the courts, that's what we do it is a litigious society' (Interview 51). And so they did, with coal companies, such as Peabody Energy, together with the COC and the NAM involved in more than 80 separate claims filed to challenge the EPA's authority (Wannier, 2010).

In January 2014, the COC and the NAM, two of the largest and most powerful industry associations, launched the Partnership for a Better Energy Future (Interviews 14 and 35). It was established to lead the 'business and industrial community in support of a unified strategy and message in response to the administration's greenhouse gas (GHG) regulatory agenda' (PBEF, 2015). According to respondents, it met 'every other week' in the offices of the COC and included Republican strategists (Interviews 4, 10 and 14). While membership was voluntary, all participants agreed to focus on the Clean Power Plan.

The Partnership for a Better Energy Future acted like a command centre tying together industry associations across almost every sector of the US economy, including in the energy, mining, manufacturing, construction, and farming sectors. In doing so, it helped to coordinate the business community's response to the Clean Power Plan by facilitating regular interaction, sponsoring studies and agreeing on strategy. Although some of these sectors would not be directly affected by the Clean Power Plan, the message from the COC and others was that coal was just the beginning.

So we are telling industry that it starts with coal, then it moves to gas, refiners, cement, agriculture.... We are Paul Revere yelling out the regulators are coming, the regulators are coming.

(Interview 10)

It was a message that resonated, with many respondents, especially in the oil and gas industries admitting that they are 'worried about the precedent, that we are next' (Interview 4).

Ahead of the EPA's announcement on 2 June 2014, which set out the proposed carbon pollution standards, business groups launched a pre-emptive strike against the regulations. The central message was that 'if the EPA regulations come to pass it will be the end of economic growth in the US' (Interview 5). The COC released analysis estimating that the EPA's plan would cost the economy $51 billion a year by 2030 and almost a quarter of a million jobs (Goldenberg, 2014). Other groups, such as the NMA, launched radio and online advertisements in key states warning of the impacts on the coal industry (Quiñones, 2014), and this was followed by a host of further studies funded by the coal, oil, and gas industries (NERA, 2014). They were supported by representatives in coal producing states, such as Republican Mike Kelly, who went as far as to compare the EPA's regulations on coal to 'terrorism' (Bravender, 2014).

According to the coal industry, the regulations represented an 'existential crisis' (Interview 35). In interviews, coal corporations and coal industry associations warned that the closure of coal plants would risk the reliability of the nation's electricity supply (Interviews 5, 6, 14, 35 and 49). The utility industry largely shared this position, including the leading electric utility association EEI. In 2014, in the midst of the cold front, or 'polar vortex', which sent temperatures to record lows and shut down parts of the US, the EEI publicly warned of the risks of closing down coal plants that provide electricity (Chediak and Weber, 2014). As one utility executive argued, '89 per cent of the coal plants that helped during the winter polar vortex are slated to go, but these plants provided the backup capacity' (Interview 6). When pressed about the claims being made by the industry, the same executive conceded that 'utilities have a habit of crying wolf, but we might be right this time' (Interview 6).

It was not just the impact of the EPA regulations on existing power plants that had the coal industry and the utility industry worried, but also what they would mean for new plants. The proposed regulations limited new plants to emissions of 499 kilograms per MWh, which in effect would require all new plants to be equipped with CCS technology (EPA, 2013). The problem for the industry as American Electric Power, one of the largest utilities in the US, pointed out is that 'there is no commercially available technology that captures carbon dioxide from an existing power plant at this time' (Bell, 2013), and those that say there is, as another argued, 'do not operate power plants' (Interview 21). In other words, 'the EPA rules mean no new coal because any new plant will need CCS' (Interview 22).

On Monday 3 August 2015, more than a year after the draft rule was proposed, the President announced the Clean Power Plan. The coal and utility industries were ready and swung into action. Within minutes of the President's announcement, West Virginia's attorney general, Patrick Morrisey, and Mike Duncan, CEO of the ACCCE, declared that a dozen Republican state attorney generals with the support of coal producers and utilities were launching a series of legal challenges to the regulations (Davenport and Hirschfeld Davis, 2015). In the coming days more would follow (E&E, 2016). The reason to take to the courts, as one respondent put it bluntly, was that 'the courts are the only game in town' (Interview 14). And coal executives explained why: 'We have tried to slow them down but to no avail, so we have to sue them over everything' (Interview 9). The challenges had been coordinated by the Partnership for a Better Future, with the COC and the NAM taking a lead role (Interviews 14 and 35).

The fight against the EPA regulations would not only take place in the courts. As Mike Duncan, CEO of ACCCE, foreshadowed back in 2013, they would also fight in state legislatures. Recalling that the Clean Power Plan provided states with significant flexibility to design and implement individual plans to reduce emissions, the coal industry attempted to enlist state actors in their fight against the regulations (Interviews 9, 14 and 35). With the financial support of utilities, such as American Electric Power, as well as the oil and gas industry, the Republican Attorneys General Association, which raised $16 million in 2014, and was led by then Attorney General Scott Pruitt of Oklahoma, an ardent critic of the EPA, helped state attorneys general to introduce legislation to obstruct the EPA (Lipton, 2014).

Many of the bills introduced were based on a model bill developed by the American Legislative Exchange Council (ALEC), a conservative not-for-profit organisation, partly funded by oil, gas, and coal interests including Chevron, ExxonMobil, and Peabody Energy (Mufson, 2015). One of the first attempts was in Kentucky where legislators passed a model bill that in effect would prevent the state from designing a state-based plan consistent with the EPA guidelines.[9] Despite the success in Kentucky, dozens of other laws failed or were watered down, in part, because utilities do not want restrictions on how they might comply with the EPA regulations (Haq, 2015). Accordingly, the Partnership for a Better Energy Future, which appeared to have been coordinating the legislative strategy, developed a new model bill that would require state legislatures to approve any state implementation plan before it is submitted to the EPA (ALEC, 2015). The effect was to open up additional avenues for business actors to lobby, because the process of designing a state plan required both the state environment agency and the legislature to support it. Pennsylvania and Arizona passed similar laws (McDonnell, 2015).

In 2015, Republican Senate majority leader Mitch McConnell gave further momentum to the opposition at the state level by developing a legal blueprint to justify why governments should not comply with the proposed regulations. Drawing on the arguments of Laurence Tribe, a Law Professor at Harvard University, who was hired by Peabody Energy to write a legal brief for the company against the EPA rules, Senator McConnell urged state governors to refuse to submit state implementation plans to the EPA. Peabody Energy has also been one of the top contributors to Senator McConnell's election campaigns (Davenport, 2014a).

Although the plan was finalised in August 2015, the Obama administration agreed to delay the submission of state implementation plans until 2018 largely in response to business opposition (Davenport, 2015). However, in February 2016 the US Supreme Court granted a stay, stopping the implementation of the Plan, which President Trump has vowed to scuttle (EPA, 2015, Adler, 2016).

Why and how did business actors shape the policy contest over the Clean Power Plan?

Preferences

The largest coal producers uniformly opposed the Clean Power Plan, aside from Rio Tinto. Coal corporations viewed the Clean Power Plan as an 'existential crisis' for the industry because it would put coal in a much less competitive position than the Waxman–Markey bill, which provided greater flexibility (Interview 35). More importantly, by, in effect, requiring all new coal-fired power plants to be equipped with CCS technology, a technology that those in the industry freely admitted did not yet exist on a commercial scale, the proposed regulations had a chilling effect by reinforcing the view in the investment community that coal is a high-risk investment (Interviews 46, 21, and 51). The culture of the industry strengthened this position. Many coal corporations continued to deride the need for action on climate change. For some, such as Murray Energy, climate change was simply a hoax (Jett, 2014). With the leading coal industry associations, namely the ACC and the NMA opposed to the Clean Power Plan, there were no signs of intra-industry conflict. Rio Tinto was again the principal exception. The reasons are likely the same. First, Rio Tinto was a much more diversified energy corporation with less than 10 per cent of its global revenues deriving from coal. Second, based in Europe, it operated in a different institutional context and like many European corporations it continued to take an active role in discussions on climate change. For example, in 2015 in the lead-up to the international climate negotiations in Paris, Rio Tinto signed the American Business Act on Climate Pledge to demonstrate its support for action on climate change (The White House 2015).

In contrast, although utilities appeared to support the Clean Power Plan, the interview data indicates that much like the Waxman–Markey bill, their preferences can be categorised into three broad groups. First, utilities that had less than 30 per cent of coal in their generation mix and growing renewable energy portfolios, such as NextEra Energy and Exelon, supported the proposed regulations because in many cases they stood to make relative gains. Second, utilities with generation portfolios that included more than 30 per cent coal continued to hedge their position because they believed that in the face of strong regulatory pressure it is better to try and influence the regulations rather than oppose them. Duke Energy took this approach as it had in the contest over the Waxman–Markey bill, despite the fact that Lynn Wood had replaced Jim Rogers as CEO (Duke Energy, 2015). The position of Xcel Energy was the same. Despite coal comprising 46 per cent of its generation capacity, it, too, sought to shape the regulations rather than oppose them so that it could gain from its renewable energy portfolio should they succeed, which had grown to almost 20 per cent by 2014 (Dunbar, 2014). Third, among the utilities that generated more than a third of their electricity from coal and stood to lose from the Clean Power Plan, Southern Company was again the principal utility to oppose the regulations outright. In comments submitted to the EPA, Southern Company claimed that:

> The proposed Clean Power Plan extends beyond EPA's authority under the Clean Air Act, is unworkable, and would increase electricity prices to customers while jeopardizing reliability. This will result in a complete deconstruction of the nation's electric sector and negatively impact America's energy security.
>
> (EPA, 2014)

While it remains unclear precisely why Southern Company took such a critical position toward the regulations, rather than hedge its position, the interview data suggests that Southern Company had close historical ties to the coal industry. And, importantly, it had also made a recent decision to invest in a CCS demonstration project in conjunction with the Department of Energy – the so-called Kemper County project, a 582 MW integrated gasification combined cycle plant, which aims to sequester 50 per cent of its CO_2 (Downie and Drahos, 2015: 9). As one utility executive pointed out, this represented a significant 'investment in the future of coal' (Interviews 46, 51).[10]

Because of the uneven distributive impact of the Clean Power Plan, some intra-industry conflict could be expected. What is interesting in this case is that at least initially there appears to have been a general disposition across the utility industry to oppose the regulations. According to interviews, there was a consensus view that the EPA rules in their initial form were simply 'unpalatable' (Interview 22). As one utility executive claimed, 'the EPA regulations have set a standard that is beyond our best plant ... you cannot build new coal plants' (Interview 6). In part, this may reflect that sections of the business community had come to prefer emissions trading, as proposed in the Waxman–Markey bill, to command and control forms of regulation, as proposed by the EPA (Meckling 2011).

Finally, it may be the case that the act of hedging works to limit intra-industry divisions and outright conflict. For example, as some utilities tried to shape the rules, the language from executives, including the EEI, softened and many executives came to support the Clean Power Plan, at least publicly (EEI, 2015). As a result, electric utilities did not break away and campaign against the EEI, as they had over the Waxman–Markey bill (Interviews 22, 6, 46 and 41). Nevertheless, the wider business environment in which utilities participated was strongly opposed to the Clean Power Plan. For example, two of the most prominent business associations, the COC and the NAM, led the campaign against the EPA in public and in the courts, and many utilities were prominent members of both these associations (Interviews 14 and 35).

Strategies

In order to shape the contest, business actors in these industries relied on a variety of strategies to shape the outcome, which in turn were dependent on the resources they had at their disposal and their political legitimacy – see Table 4.2. First, firms mobilised coalitions and leveraged diverse networks. While the coal and utility industries did not mobilise significant coalitions to support the regulations, those opposed did. An ad hoc coalition was established, the Partnership for a Better Energy Future, as the command centre for the campaign. In contrast to USCAP, which had acted to coordinate business actors to shape the rules around emissions trading, the sole purpose of the Partnership was to lead 'a unified strategy' against the EPA regulations (PBEF, 2015).

> The Partnership aims to educate and mobilize the broader business community and elected and public officials to address widespread concerns with forthcoming greenhouse gas rules.
>
> (PBEF, 2015)

Established by the COC and the NAM, it tied together more than 200 coalitions including state and national associations from the mining, manufacturing, transport, farming, oil, and gas sectors, not to mention dozens of state chambers of commerce (PBEF, 2015). In doing so, it was able to concentrate members' resources to pursue the common goal of defeating the federal regulations. As one member explained:

> It is just informal … we just get together now and again and divvy up tasks. If it's too formal it's too hard to get stuff done. Being nimble has its advantages, especially when you have the complexity of 50 states.
>
> (Interview 23)

Table 4.2 Summary of the policy contest over the Clean Power Plan

	Coal producers	Electric utilities
General industry preferences	Oppose	Support/hedge/oppose
Key coalitions mobilised in the energy sector	ACCCE, ACC, NMA	EEI
Key business coalitions mobilised	Partnership for a Better Energy Future, COC and NAM	Partnership for a Better Energy Future, COC and NAM
Other important actors in the networks	Republican Attorneys General Association	Republican Attorneys General Association
Framing strategy	Framed around economic costs and job losses	Framed around economic costs and job losses
Lobbying	The coal industry spent $18 million on lobbying between 2014 and 2015	The utility industry spent $240 million on lobbying between 2014 and 2015
Outcome	Clean Power Plan delayed	

In other words, it derived its power not from its own resources, but from its ability to tie multiple coalitions together and leverage diverse networks, including state actors. For example, the Partnership, with the support of coal producers, such as Peabody Energy, and utilities, such as American Electric Power, enrolled the Republican Attorneys General Association in opposition to the Clean Power Plan, which in turn introduced bills into state legislatures that made it difficult for state governments to comply with the EPA regulations. This, of course, was assisted by direct lobbying and campaign contributions, which helped to open doors.

A second key strategy was framing. The Partnership used its legitimacy derived from its breadth of business members to frame the contest around economic costs and job losses. As Mike Duncan from the ACCCE claimed, the Clean Power Plan will result in a 'shattered economy' (Duncan, 2013). Recognising the strategic importance of information for framing, these coal coalitions and their members commissioned studies, which showed millions of job losses if the regulations came to pass. They also linked the Clean Power Plan to energy security exploiting extreme weather events, such as the polar vortex in 2014, to warn of the risks of targeting coal-fired power plants. Because business actors did not mobilise significant coalitions in support of the EPA regulations there were few instances of counter-frames, at least within the business community, though as previous studies have shown business actors in support of climate change tend to frame regulations as an economic opportunity rather than a threat (Jones and Levy, 2007). This was certainly the approach taken by USCAP in advocating for emissions trading (USCAP, 2009).

Third, business actors continued to lobby, but the declining financial resources of the coal industry meant that it did not match the scale of the Waxman–Markey bill. As one coal lobbyist quipped, compared to past regulatory battles 'coal is so strapped for cash at the moment ... they are not doing anything federally' (Interview 49). For example, between 2014 and 2015 coal producers spent $18 million on lobbying, compared to $34 million between 2009 and 2010. Electric utilities, on the other hand, were in a much stronger position spending $240 million between 2014 and 2015 (CRP, 2014, 2016e).

However, it was not just financial resources that were frustrating the Partnership's lobbying efforts, it was also that their lobbying pathways were limited. Unlike the Waxman–Markey bill, the Clean Power Plan was being implemented via regulations and not legislation. As one executive explained, 'there are more avenues to lobby on a bill, but there is much less on a regulation because you can only lobby the administration' (Interview 14). And as another coal lobbyist claimed, 'this administration is not open to discussion' (Interview 49).

As a result, business actors engaged in what in scholars describe as 'forum-shifting' (Braithwaite and Drahos, 2000) or 'venue shopping' (Baumgartner and Jones, 1991), which refers to how actors take actions in different forums to influence governance outcomes. In other words, when lobbying in one forum failed, business actors simply tried another. For example, with little hope of blocking the regulations in Congress, coal companies turned to state legislatures to prevent state governments from complying with the EPA regulations, and they turned to the courts. Indeed, the courts proved to be a particularly significant forum for business actors given the adversarial nature of the US legal system (Kagan, 1991). As discussed, coalitions of coal producers, electric utilities, and industry associations often in conjunction with state governments, such as West Virginia, filed more than 100 separate claims challenging the EPA's authority to regulate greenhouse gas emissions. Duke Energy and Xcel Energy, for example, which publicly supported the regulations, did not dissociate from groups that were fighting the regulations in court or working with state governments to do so, such as the Utility Air Regulatory Group (Walke, 2012). Indeed, both had funded law firms that worked in the courts to undermine EPA regulations and they were by no means the only ones. For example, American Electric Power continued to do the same, as one insider explained, 'the vast majority [of utilities] are not suing the EPA, but that does not mean they are not pushing state governments to do so on their behalf' (Interview 41 and 6). In most cases they were defeated, but they did succeed in delaying the Clean Power Plan and creating a sense of uncertainty around the regulations.[11]

Finally, what influence did business actors have on the policy contest? In contrast to the Waxman–Markey bill, there were fewer business actors in the energy sector in support of the Clean Power Plan, and fewer still appear to have actively campaigned for it. Instead the most active firms, especially in the coal and utility industries, were opposed and their coordinated resistance campaign helped to slow the implementation of the Clean Power Plan. Yet, once again, the mobilisation of other non-state actors and the role of policymakers helped to create and limit opportunities for business actors to shape the contest. First, while not the focus of this analysis, environmental NGOs did mobilise not only to support the Clean Power Plan, but also to help put it on the agenda. The NRDC in particular was instrumental in shaping and advancing the EPA draft rule. This was later used by Congressional Republicans as grounds to oppose the plan and investigate the EPA in 2014 (Davenport, 2014b). As a result, for coal producers and electric utilities opposed to the Plan, the mobilisation of other non-state actors likely limited their influence especially with the Obama administration and Democrats in Congress that worked closely with organisations like NRDC.

Second, and related, because Republicans in Congress and in state legislatures overwhelmingly opposed the Plan there were opportunities for business actors both to reinforce their opposition, through lobbying, for example, but also to incorporate them into coalitions and networks in ways that could frustrate the Obama administration. This was especially evident with the role played by the Republican Attorneys General Association at the state level. The political incentives for many Republicans were reasonably straightforward. The Republican leadership had made a strategic decision to oppose the President's climate plans irrespective of the detail and individual representatives and senators, especially from coal producing state like West Virginia, Pennsylvania, and Kentucky, were well rewarded by coal producers and utilities for their opposition to the regulations. Of course, this was not the case in every state, and Democrats often had public polling on their side in making the case for climate action. However, in the contest over the Clean Power Plan, there were very few firms among the incumbent fossil fuel industries that were willing to support them, and in addition, all the major business associations, such as the COC and the NAM, were opposed.

Conclusion

The so-called war on coal has many battlefronts. During the Obama administration, two of the most important have been the contests over the Waxman–Markey bill and the Clean Power Plan. Business actors in the coal and utility industries, as well as others right across the business community have played a significant role in shaping these policy contests. These actors merit attention. To be sure, coal is the largest source of greenhouse gas emissions in the world and the US still holds the world's largest coal reserves on the planet. Yet any efforts to restrict emissions from coal will not succeed unless they overcome the resistance of incumbent industries in the coal industry and the utility industry that have generated great wealth from burning coal. As such, these policy contests provide excellent cases to examine how and why business actors in US energy sector behave.

Tables 4.1 and 4.2 summarise business behaviour in both policy contests. In the coal industry all but one of the top coal producers opposed emissions trading and direct regulation because of the adverse impact they would have on their competitive position in the energy market. In both cases coal producers stood to lose given that most corporations relied on coal production for close to 100 per cent of their revenues. The exception was Rio Tinto, which was a much more diversified energy corporation headquartered in the UK, not the US. This meant that it operated in a different environment, one which was generally more supportive of action on climate change.

In the utility industry the data reveals more nuanced positions, which were closely correlated to generation portfolios. Specifically, electric utilities can be categorised into three broad groups that supported, hedged or opposed the regulatory initiatives. Hedging was most evident in the utility industry and manifest in two ways, namely working to shape regulations, especially when they looked likely to succeed, and/or supporting them publicly while indirectly opposing them. This is consistent with recent work on firm preferences in global environmental politics, which suggests that hedging strategies are an especially prevalent form of corporate behaviour (Meckling 2015). In addition, the uneven distribution of preferences across the utility industry did result in intra-industry conflict, primarily in the contest over the Waxman–Markey bill. While the leading industry association, the EEI, supported emissions trading, other EEI members broke ranks and campaigned against their industry colleagues. Such divisions were much less apparent in the contest over the Clean Power Plan as many utilities that initially opposed the regulations came to support them, at least publicly, as they looked more likely to succeed.

In pursuing their preferences business actors mobilised coalitions and leveraged networks to shape the outcome. In particular, key coalitions pooled resources and coordinated activities. In the contest over emissions trading USCAP played this role tying together networks in the coal, oil and gas industries, not to mention NGOs, in support of the Waxman–Markey bill. Whereas the ACCCE led the opposition from the coal industry and sections of the utility industry, the Partnership for a Better Energy Future played this role in response to the Clean Power Plan tying together more than 200 coalitions across multiple industries. Consistent with the last chapter, the creation of the Partnership highlights the continued appeal of ad hoc coalitions that are temporary in nature and issue-specific. These coalitions in turn leveraged their diverse networks to enlist other firms across the business community, such as in the oil and gas industries, as well as state actors at the sub-national level, notably the Republican Attorneys General Association, which worked alongside industry to oppose the regulations both in state legislatures and in the courts. These networks were also built to improve legitimacy. For example, in the contest over emissions trading, electric utilities leveraged their historical ties to the union movement to carry sway with Democrats in the Congress (Interview 6).

In the coal and utility industries framing was also a critical strategy. In both cases these industries defined the proposed climate regulations as the problem and framed the contest around economic costs and job losses. For example, in the contest over the Waxman–Markey bill, coal producers drew on their discursive power to consistently label emissions trading as an economic threat highlighting the job losses that would flow from its implementation (Peabody Energy, 2014). And they strategically deployed the findings of economic studies to reinforce these frames. Key coalitions took the lead in this process as they did in coordinating much of the lobbying activity, which was supported by considerable financial resources. After all, coal producers and utilities spent hundreds of millions on lobbying to kill and to shape the policy outcomes. Yet for the coal industry at least, their declining financial resources inhibited their lobbying activity in response to the Clean Power Plan, as did the more limited pathways for lobbying directed at regulations compared to legislation.

Finally, these policy contests, like those in the oil and gas industries, highlight the capacity of business actors in the incumbent fossil fuel industries to shape governance outcomes. For example, it appears likely that the support of key coalitions and firms in the utility industry for the Waxman–Markey bill, which were directly impacted by the emissions trading scheme, were vital to ensuring the passage of the bill through the House, even if it failed in the Senate. Likewise, there is little doubt that the campaign mounted by actors across the business community, but especially in the energy sector, in tandem with Republican opposition at the federal and state level, helped to delay and ultimately derail the implementation of the Clean Power Plan.

Yet, as evident in the preceding pages, the influence of business actors was context dependent. The capacity of firms in these industries to impact outcomes was conditioned by the mobilisation of other non-state actors and the role played by policymakers, who in turn were driven by their beliefs and political incentives. Indeed, in both cases the mobilisation of environmental NGOs created and limited opportunities for business actors to shape outcomes. For example, business actors, especially in the utility industry, cooperated directly with environmental NGOs via USCAP to enhance their capacity to shape the design of the Waxman–Markey bill. In contrast, in the contest over the Clean Power Plan, there appears to have been little cooperation between environmental NGOs, which had helped to put the Plan on the agenda, and the coal industry and the utility industry that were actively opposing it. In other words, the mobilisation of non-state actors, such as NRDC, likely limited capacity to influence the contest.

Equally important was the role of policymakers. The fact that the contests over the Waxman–Markey bill and the Clean Power Plan took place at all was the result of the President's belief in climate change action and the belief of many in his party, including congressional leaders. There were also political incentives to act with national opinion polling showing support for action, especially among Democrats, as well as support from a range of other non-state actors. This is one of the reasons that some firms in the coal and utility industries worked to hedge their positions because they saw the writing on the wall. For those that decided to oppose however, they were often able to enlist Republicans to their cause, who typically were ideologically opposed to climate change measures and or, had strong domestic incentives in their home states to oppose, because they were coal-producing states and or, states where coal was a major source of electricity generation. Further, while polling may have shown support for climate change, this was tempered by the fact that it rarely ranked as a high priority issue for voters (Egan and Mullin, 2017). Business actors exploited these state-based incentives through outside lobbying campaigns, which raised the salience of the issue and brought constituent pressure to bear, including on Democrat representatives who on occasions decided to oppose the White House. Of course, in the case of the Clean Power Plan, such campaigns were more limited given the White House had turned to regulation rather than legislation, but Republicans at the state level provided useful allies too.

In summary, the focus on business actors in these industries is revealing for both scholars and policymakers because it highlights the nuanced preferences within and between industries, especially among those actors that hedge their position, and the complex means by which they pool resources and coordinate activities to exercise influence over the policy process. Once again, the industry divisions, combined in these cases with the structural decline of coal, presents a fractured industrial landscape that offers opportunities and pitfalls for policymakers seeking to advance a clean energy transition.

Notes

1 Sections of this chapter have been previously published, see Downie, Christian. 2017b. Fighting for King Coal's crown, *Global Environmental Politics*, 17:1, 21–39. © 2017 by the Massachusetts Institute of Technology, reprinted courtesy of the MIT Press.
2 Interview 6.
3 Interview 6.
4 The EPA began experimenting with domestic emissions trading with the introduction of the Emissions Trading Program in 1974. Other programs followed with the Lead Trading Program in 1982 and chlorofluorocarbon trading several years later (Fisher-Vanden, 2000).
5 USCAP members included: AIG, Alcan, Alcoa, BP America, Boston Scientific, Caterpillar Inc., Chrysler, ConocoPhillips, Deere & Company, Dow Chemical, Duke Energy, DuPont, Environmental Defense, Ford Motor Company, FPL Group, General Electric, General Motors, Johnson & Johnson, Lehman Brothers, Marsh, Natural Resources Defense Council, PepsiCo, Pew Center on Global Climate Change, PG&E Corporation, PNM Resources, Shell, Siemens, The National Wildlife Federation, The Nature Conservancy and World Resources Institute.
6 Interview 35.
7 In *Massachusetts* v. *EPA* (2007), the Supreme Court ruled that 'air pollutants' under the Clean Air Act encompass greenhouse gases. The Supreme Court upheld the EPA's authority in subsequent rulings. In 2009, the EPA issued an 'endangerment finding' that greenhouse gases endanger public health, providing the basis for the EPA to regulate greenhouse emissions (EPA, 2009a).
8 States are permitted to reduce emissions via four 'building blocks': increasing the efficiency of fossil fuels power plants; (ii) increasing the capacity of low emission plants, such as natural gas; (iii) use zero-emission sources, such as nuclear power; and (iv) increase electricity efficiently.
9 The bill excluded the Kentucky Energy and Environment Cabinet, the agency responsible, from using the building blocks set out by the EPA, such as renewable energy or energy efficiency, to reduce emissions (Chemnick, 2015).
10 Despite favourable funding conditions, including support from the DoE, the project has experienced numerous delays and cost blow-outs and in 2017 it was officially abandoned (Chediak, 2017).
11 Some business groups, such as the Business Roundtable, decided not to participate in legal challenges to the EPA regulations.

5 The rise of renewable power
Policy contests in the wind and solar industries

Introduction

> Solar is going to be the future.… It is just a matter of tearing down the barriers to allow that to happen.[1]

Around the world renewable energy is booming. According to the IEA, it could be the world's largest source of electricity supply before 2030 (IEA, 2016b: 397). The US is now adding more new renewable energy capacity than fossil fuel capacity and this is expected to continue in the coming decades (EIA, 2016g). This has been driven both by technology improvements and government incentives, which have seen renewable energy, especially wind and solar power, become increasingly competitive in recent years. The hype around renewable energy also reflects the development of battery storage technology. Since Tesla announced its Powerwall in 2015, a rechargeable lithium-ion battery to store electricity, growing attention has focussed on the capacity for renewable energy to provide base load electricity at a cost at which none of the existing energy sources can compete. As almost all respondents agreed, 'solar storage will completely change the game' (Interview 22), or as one oil industry representative put it, 'if someone figures out battery technology, my industry will be in trouble' (Interview 42).

While the future of renewables looks bright, globally wind and solar power still account for less than 1.5 per cent of total primary energy supply (IEA, 2017a: 8). Government policy will have a significant impact on how bright the future will be. The history of renewable energy policy in the US has been anything but consistent. Since President Carter put solar panels on the White House in 1977 and increased subsidies, the sector has won and lost a series of regulatory battles (Laird and Stefes, 2009: 2620). In the 1980s, President Reagan removed many of those same subsidies, including research and development funding, tax credits and accelerated depreciation. History was largely repeated in the 1990s when tax credits for renewable technologies that were established in 1992 in the aftermath of the first Gulf War expired in 1999 (Sovacool, 2009: 4507–4508).

There are many similarities today. In this chapter the focus is on the wind and solar industries because of all the renewable sources that are commercially viable, these two have the greatest potential to transform the energy sector (IEA, 2015b). It also means that these are the industries where some of the fiercest policy contests are taking place between the traditional incumbents and their rivals in the renewables sector. In order to consider how and why business actors are shaping the rules that govern wind and solar power, this chapter will focus on two contemporary policy contests: the first is the battle to retain the federal production tax credit (PTC), which has encouraged the development of the wind industry; and second is the contest over the federal investment tax credit (ITC), which has facilitated the development of the solar industry. What is striking about these policy contests is that, crudely speaking, they pit the incumbent industries, especially the utility industry, against the new kids on the block, the wind and solar industries.

This chapter follows the structure of the last. The next section provides an overview of the US renewable industries, including the business actors in the wind and solar industries. The following sections analyse the behaviour of business actors in the contest over the PTC and the contest over the ITC.

Overview of the US renewable industries

Renewable energy is flourishing around the world. In 2015, capacity additions of solar, wind and hydro surpassed those of fossil fuels and nuclear energy for the first time and world renewables based capacity generation now exceeds that of coal, even if it is yet to produce as much electricity (IEA, 2017b). In the US, renewable energy is also on the rise and in 2017 renewables accounted for 12.7 per cent total US energy consumption (EIA, 2018h). This represents the highest renewable energy share since the 1930s, when wood was a large contributor to domestic energy supply (EIA, 2015c). Renewable energy also increased to equal 17 per cent of US utility-scale electricity generation. Hydropower (7 per cent) and wind power (6 per cent) comprised the majority of renewable generation with solar power producing around 1 per cent (EIA, 2018e).

First, wind, which along with solar power, has led the recent growth in renewables based capacity, accounting for 94 per cent of total growth in the reference case – see Figure 5.1 (EIA, 2018a: 94). The growth in wind power is the result of both improvements in wind technology and policy measures including the tax incentives for renewable energy, which are the focus of the policy contests below (EIA, 2016h). In the EIA's reference case, wind generation is projected to increase rapidly in the near term to take advantage of existing tax credits before they expire, and then rises more modestly after 2025, increasing 20 GW from 2020 to 2050 (EIA, 2018a: 96).

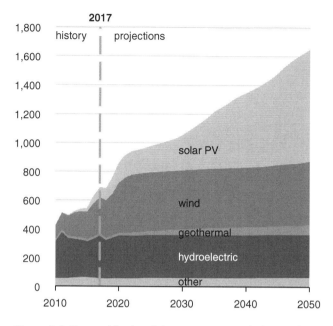

Figure 5.1 Renewable electricity generation, including end-use
    generation, reference case.

Source: Energy Information Administration (EIA, 2018a).

Second, solar power has also been surging in recent years. Since 2011,
solar generation in the US has increased by more than 2,000 per cent (IEA,
2014b). Utility-scale solar installations, which include PV and thermal technologies, increased by more than 70 per cent per year between 2010 and
2016. Small-scale solar systems, such as rooftop solar, have also grown (EIA,
2017g). This has been largely driven by a dramatic decline in the cost of PV.
Between 2008 and 2014, in the US the average price of PV modules fell by
84 per cent from $4.00 per peak watt to around $0.65 per peak watt (MIT,
2015: 79). Notwithstanding, the solar share of utility-scale electricity generation has not passed 1 per cent. However, the EIA projects this to change in
its reference case – see Figure 5.1. Both utility-scale solar and distributed
solar are expected to rise to 14 per cent by 2050 with electric generation
from utility-scale solar closely followed by distributed solar (EIA, 2018a: 98).
While utility-scale solar generally has lower costs, distributed solar often
competes against higher retail electricity prices.

While wind and solar power are typically the focus of renewable energy
discussions, in the context of climate change nuclear power has also been
put forward as an alternative to fossil fuels because it is a low carbon fuel
source. However, most energy scenarios show nuclear power declining, in

direct contrast to wind and solar power. In the US, nuclear power comprises 9 per cent of generation capacity, though in 2016 it contributed almost 20 per cent of total utility-scale electricity generation because it has higher capacity factors than any other generating technology. Nevertheless, the EIA projects that the nuclear share of US electricity generation will fall from 20 per cent in 2016 to 11 per cent in 2050. In short, it will almost halve in the coming decades. This reflects the likelihood that more nuclear capacity will be retired than built given the aging nuclear fleet. In fact, the vast majority of the 99 nuclear reactors operating across 61 plants in the US as of 2016 came online between 1970 and 1990. In addition, utilities have been shelving construction plans on new reactors because of the deteriorating economic case for nuclear power (Adcox, 2017; EIA, 2017c).

The rise of renewable energy is having widespread impacts across energy markets. In the US, this is most evident in the electricity market where wind power, and particularly solar power, are threatening the business model of electric utilities, which is why, as I will discuss below, electric utilities are resisting. There are three principal reasons. First, utilities cannot profit from renewable assets they do not own. Traditionally, utilities develop new electricity capacity and in return state regulatory commissions guarantee a rate of return to cover the costs of these projects. Yet in the case of distributed solar, for example, where households and solar companies are building the new capacity, utilities cannot apply for a guaranteed rate of return. Second, the uptake in distributed solar is disrupting the utilities monopoly, and with it, its profits. As distributed solar becomes more competitive, it not only reduces the amount of electricity utilities can sell to their consumers, but it also means utilities must spread the fixed costs of operating the grid over fewer kWh. As a result, utilities are forced to increase their rates to recover the costs making solar more competitive still; the so-called 'death spiral' for utilities. Third, in competitive electricity markets wind and solar power put downward pressure on the wholesale price of electricity. In the case of wind, government incentive schemes that encourage wind operators to produce power irrespective of the price, such as the PTC, reinforce the pressure on prices. This can lead renewable operators to produce power at prices below zero, which is not uncommon in US energy markets. In other words, wind operators may choose to pay buyers to take the power rather than reduce their output when demand is insufficient. The net result is lower prices. Similarly, because solar operates when the sun is shining, the peak time for electricity use, it reduces the number of hours for peaking plants, which operate at high prices for a few hours each day to meet peak demand. As a result, it lowers the price of these peak hours, which in turn reduces utility profits and can lead to the early retirement of fossil fuel plants (Stokes, 2015: 28–30; Satchwell *et al.*, 2015).

Although the growth in renewable energy is already impacting conventional energy markets, it will need to accelerate to be consistent with the 2°C scenario. As the IEA makes clear 'any credible path to achieving the world's climate objectives must have renewable energy at its core' (IEA, 2016b: 398). Indeed under the 450 Scenario nearly 60 per cent of the power generated worldwide in 2040 is projected to come from renewables, almost half of which is from wind and solar power – see Figure 5.2 (IEA, 2016b: 412). This significant scale-up in wind and solar capacity reflects the cost reductions, widespread availability and their relatively short construction times. For example, by 2040 the global average capital costs of solar PV are less than half their current level (IEA, 2016b: 430). In the US, renewable energy becomes the largest source of generation by around 2035. While electricity demand grows slowly at 0.5 per cent per year to 2040, most of the growth comes from the increase use of electric vehicles (IEA, 2016b: 527). The effect of this transformation is a sharp decrease in energy-related emissions, which decouple from demand. Globally, by 2040 energy demand is less than 10 per cent higher than current levels, but emissions are 43 per cent lower, largely because of cuts from the power sector (IEA, 2016b: 429).

So, who are the business actors in the wind and solar industries? First, the wind industry essentially comprises two groups of actors: wind manufacturers and wind power operators. Wind manufacturers make utility-scale wind turbine blades, towers, generators, and other related components. During the period of the policy contests, the big three wind manufacturers operating in the US were GE, Vestas, and Siemens. In 2015 GE captured 40 per cent of the wind power market followed by Vestas with 33 per cent and Siemens with 14 per cent. Other companies include

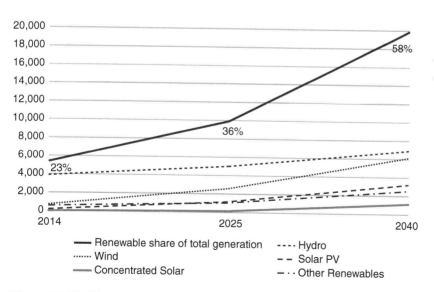

Figure 5.2 World Renewables Consumption (TWh) under 450 Scenario.

Source: International Energy Agency (IEA, 2016b: 412).

Acciona, Gamesa, and Nordex. While most of these firms own manufacturing facilities in the US, significant competitive pressure on global supply chains has been part of the reason for a number of mergers in the industry, including between the German company Nordex and the Spanish company Acciona in 2016 (DoE, 2016).

The wind power segment of the industry generates revenue from owning and operating wind farms and selling the power produced downstream. Independent power producers (IPPs) own the vast majority of wind assets in the US. In 2015 IPPs owned 85 per cent of new wind capacity, utilities 12 per cent, with 2 per cent owned by other entities, such as towns, businesses, and farms (DoE, 2016). The top five wind project asset owners – NextEra Energy, Iberdrola Renewables, Berkshire Hathaway Energy, EDP Renewables North America, and Invenergy – own around 40 per cent of the installed wind power capacity in the US (AWEA, 2014). Most of the wind power produced is sold via PPAs to utilities, though direct retail sales to businesses have grown in recent years with large companies, such as Facebook, Google, and Ikea purchasing wind power directly from generators rather than from utilities (DoE, 2016). Among utilities, Xcel Energy is the top provider of wind energy followed by Berkshire Hathaway Energy, Southern California Edison, and American Electric Power, though for a number of these utilities wind is only a small part of their generating mix (AWEA, 2015a). More than 50 per cent of US wind capacity is located in five states – Texas, Iowa, Oklahoma, California, and Kansas – and Texas alone accounts for almost a quarter of total US wind capacity (EIA, 2017h).

Second, the solar industry in the US broadly comprises two principal segments: solar manufacturers and installers that focus on large-scale or utility-scale solar; and solar installers with some manufacturing capacity that focus on small-scale or distributed generation solar (Osten, 2015; Khedr, 2015). Some electric utilities own and operate solar power generation as well. Although solar is booming in the US, the growth has been uneven. The industry has witnessed bankruptcies and consolidation as companies strive for lower costs in the face of competition from both domestic and international corporations.

In the period of the policy contests, the German-based Solar World and American-based First Solar were the largest solar manufacturers and the largest solar firms in the US. Together, they represent over a third of the solar manufacturing market in the US (Osten, 2015: 28–30). In 2014, First Solar's annual revenues were over $3 billion, and it is responsible for some of the largest solar plants in the US, including Topaz Solar and Desert Sunlight in California, which went online in 2014 and 2015, both with a generation capacity of 550 MW (REN21, 2015: 61; Osten, 2015: 29). These solar plants are commonly referred to as utility-scale solar because the power they generate feeds directly into the electricity grid and is typically purchased by a utility company via a PPA. Other large manufacturing firms that operate in the US include Suniva, Trina Solar, and Yingli Solar.

In the solar installation segment of the industry, which is highly fragmented, many companies are small operators. Over the same period, the largest firms, some of which have a manufacturing capacity, include SolarCity and Sunpower. In 2014, these two companies represented about 7 per cent of the solar installation market. And SolarCity, for example, had revenues totalling $255 million (Khedr, 2015: 23). Other notable firms include Vivint Solar, Sungevity, and Sunrun. SunEdison, another large player in the solar market, filed for bankruptcy in 2016 (Roselund, 2016). Although utility-scale solar dominates the industry, the retail installation market has experienced significant growth. Distributed generation solar, as it is commonly referred to, is produced at or near the point where it is used, mainly on residential or commercial roofs. While there are a range of competing solar technologies, PV systems in the US account for 90 per cent of installed capacity split between utility-scale plants and residential and commercial installations (MIT, 2015: xv).

Like its counterparts in the oil, gas, coal, and utilities industries, wind and solar power have a range of industry associations. The American Wind Energy Association (AWEA) is the largest and most prominent association for the wind industry, though there are also a range of smaller associations, which are often state-based, such as the California Wind Energy Association. The Solar Energy Industry Association (SEIA) is the most active in the solar industry. There are also associations representing particular segments of the industry, such as The Alliance For Solar Choice (TASC), which was created by SolarCity and others to advocate for rooftop solar, and the Large-Scale Solar Association (LSA), which focuses on utility-scale solar. The wind and solar industries are also represented in renewable energy groups, such as the American Council on Renewable Energy (ACORE), and other advocacy groups, such as Advanced Energy Economy (AEE).

The policy contest over the PTC: 'If you're a wind company you support it'[2]

The US energy sector has a long history of government subsidies. Fossil fuel companies receive billions in subsidies every year and while the sums are smaller, the renewable sector is subsidised as well. One of the principal subsidies for wind power is the PTC: an inflation adjusted tax credit for electricity generated by qualified sources, such as wind power. Initially established under the 1992 *Energy Policy Act*, the PTC reduces the price of wind power thereby making it more competitive with others forms of electricity. However, the PTC has been renewed and revised numerous times and the constant uncertainty over its renewal has produced a boom-bust pattern of investment (Barradale, 2010). As many in the industry were quick to point out, 'the fits and starts' of the PTC had damaged the wind industry and the uncertainty had resulted in sharp declines in the amount of wind power installed (Interview 53).

While the policy contest over the PTC has outlasted several presidents, the focus here is on the most recent battles in President Obama's second term (2013–2017). Indeed, the second inauguration of President Obama in January 2013 was followed by a renewed push by the wind industry – both wind manufacturers and wind developers – to extend the PTC, which had been extended as part of the *American Taxpayer Relief Act of 2012*, but was due to expire again at the end of the year (DoE, 2016). With the support of the White House, Democrats, and a number Republicans the wind industry had some success in extending the PTC, but they faced significant opposition from business actors in the energy sector, especially some sections of the utility industry, and Republicans in Congress. The unlikely deal reached in December 2015, which was closely tied to the fate of the solar ITC, extends the PTC to 2019. The tax credit is set to phase down in 20 per cent year increments with projects constructed in 2017 receiving 80 per cent of the PTC value, 60 per cent in 2018 and 40 per cent in 2019 (DoE, 2016). In 2015, the PTC for wind was $23 per MWh (DoE, n.d.).

The AWEA was the 'key architect' of the campaign and as the largest wind industry association in the country it had both the resources and the legitimacy to carry it out (Interviews 54 and 56). For example, between 2013 and 2016, the period of the of policy contests examined here, it spent around $3.5 million on lobbying (CRP, 2015a). It also had close ties to policymakers in the White House and Congress (Interviews 54 and 52). However, the support across the industry for the PTC did not manifest in a uniform position and this caused problems for the AWEA. A former senate staffer explains the history:

> In the early stages there was alignment across the wind industry, but there was a breakdown around 2011 and 2012. The 2010 election wasn't so great for Democrats and so it was in 2011 and 2012 that people went, 'oh boy we don't know how many extensions we can get so we need new political approaches'.
>
> (Interview 54)

This led to a split within the industry. According to industry representatives, there were two camps – those companies that supported a phase-out of the PTC and those that supported a multi-year extension (Interviews 54, 55, and 56). The principal firms supporting a quick phase-out appear to have been NextEra Energy, the largest wind asset owner in the US, and GE, the largest wind manufacturer in the US. In the second camp were wind manufacturers and wind developers, such as Iberdrola Renewables, EDP Renewables, Pattern Energy, and Vesta, which had their own ad hoc group, discussed below (Interviews 54 and 55).[3] The head of Siemens USA, Eric Spiegel, argued at the time that the industry needs a policy that 'goes on for five or six years, because this kind of starting and stopping really hinders the market' (Fifield, 2013). Although 'AWEA did not take a side', this did hamper their capacity to lobby on behalf of the industry as the policy contest progressed (Interview 55).

In April 2013, the AWEA announced Tom Kiernan as their new CEO and his primary task was to ensure that the PTC would be extended (AWEA, 2013). He was not alone. While other incumbent fossil fuel industries, especially the utility industry, did not support tax credits for renewable energy, as I will discuss, there was a broad network of actors that did, including in the investment community and the environmental community. For example, CERES, a Boston-based NGO, founded the Business for Innovative Climate and Energy and Policy (BICEP), whose members, which included firms such as Mars, Starbucks, and Gap, lobbied Congress to extend the PTC (Plautz, 2013; Ceres Policy Network, n.d.). This helped the wind industry not only by expanding their financial resources with the support of the investor community, but also by enhancing the legitimacy of the campaign given the support from business actors not traditionally associated with the wind industry.

As the expiration of the PTC drew closer, the policy contest heated up. In October 2013, the House Subcommittee on Energy Policy, Health Care, and Entitlements held a congressional hearing on the PTC, which coincided with the launch of a campaign to end the PTC by a network of conservative political groups, such as Americans for Prosperity, which is funded by the Koch brothers and has traditionally supported fossil fuels. These groups were supported by Republicans in Congress. For example, in November 51 Republicans in the House of Representatives, led by Representative Mike Pompeo, urged the House Ways and Means Committee to allow the PTC for wind to expire. In response the AWEA worked to mobilise its own coalitions, such as the Governors' Wind Energy Coalition, which wrote to the Congressional leadership advocating for a multiyear extension of the tax credit (Copley, 2013). Despite the efforts of the industry, the PTC expired at the end of 2013.

In 2014, the AWEA continued to lead the charge by working to have the PTC renewed as part of a broader package of tax extenders. As Tom Kiernan made clear, the focus initially was on the Senate Finance Committee where the industry had strong support from traditional advocates, such as Republican Senator Grassley, who had first sponsored the PTC in 1992.[4] However, by mid-2014 negotiations in the Senate had been commandeered by Senator Majority Leader Harry Reid's office (Montgomery, 2014). The industry and its supporters in Congress framed the contest around jobs and growth. For example, the AWEA claimed that the extension of the PTC would ensure that 'tens of thousands of workers in US wind manufacturing facilities' could get back to work (Whitlock, 2013). Paralleling the approach of business actors in other contests, the wind industry sought to reinforce its framing efforts with the strategic use of information, for example, by releasing annual reports highlighting the number of jobs supported by the industry (AWEA, 2014).

Although the industry succeeded in the Senate Finance Committee, they did not have it all their own way. Some actors in the utility industry, with the support of a wider network of conservative groups, pushed back against efforts to extend the PTC. While the largest coalition of utilities, EEI, 'stayed quiet', in part because their members were split on the issue, a number of utilities with nuclear portfolios became vocal opponents (Interview 52). Exelon, the largest nuclear reactor owner in the US, was the leader and it had the financial resources to back it up. Between 2013 and 2016 Exelon spent $14.8 million on lobbying. While these financial resources were not devoted entirely to the contest over the PTC, they far outweighed the sums marshalled by the AWEA (CRP, n.d.). Exelon argued that the PTC is distorting electricity markets by enabling wind generators to sell into wholesale electricity markets at negative prices thereby driving out nuclear power (Trabish, 2014). Its position was one of the reasons 'they were thrown out of the AWEA' a few years earlier (Interview 55). The AWEA was quick to hit back, releasing a report that sought to debunk the claims being made, and it was joined by other utilities including NextEra Energy and Xcel Energy (Trabish, 2014).

> Exelon, the largest owner of merchant fossil and nuclear power plants in the U.S., has been leading a campaign to undermine the broad support for wind energy with the argument that the lower electricity prices brought about by wind energy are somehow a bad thing.
>
> (Goggin, 2014: 2)

Putting aside the technical debates around negative pricing for the moment, the aim of Exelon, according to a number of respondents, was simply to protect their bottom line. As one former senate staffer put it: 'they were in trouble and they figured it was easier to blame wind than natural gas' (Interview 54). This may have reflected the fact that coalitions associated with the oil and gas industry, such as the API, were also opposed to renewable tax credits (Interview 8). And they were joined by a network of conservative lobby groups, which continued to be led by Americans for Prosperity. For example, in June 2014, the group published a full page advertisement in Politico attacking the wind industry, which included the signatories of related groups, such as the American Legislative Exchange Council that also received funding from the Koch brothers, and oil and gas companies, such as ExxonMobil (Elsner, 2014).

Ultimately, these efforts reinforced Republican opposition and contributed to the collapse of a deal on tax extensions. Instead, in December 2014 Congress passed a fall back option, the Tax Increase Prevention Act of 2014, which extended the PTC as part of package of more than 50 other tax credits that had expired at the end of 2013 (Montgomery, 2014). The Act enabled the wind industry to claim the PTC retrospectively for

2014, but it was far from what the industry had wanted. As a result, the AWEA was quick to gear up again. In a December 2014 fundraising email it called on members to join the fight in January for policies that will advance clean energy (Whieldon, 2014).

And fight is what they did. In 2015 the AWEA went on the offensive as it sought to mobilise actors in support of the PTC. At the annual wind power conference, Tom Kiernan told firms in the industry that 'we need to build the political power of this industry commensurate with our size and scale', ramping up the rhetoric in a manner common in these contests, he argued that the industry needs to be 'publicly chastising those politicians opposed to the PTC and 'making sure they're not re-elected' (Copley, 2015a). Lobbyists spoke of the emphasis on the 'grassroots and the grasstops' (Interview 52). In other words, the industry targeted thousands of their members (grassroots) as well as opinion leaders in different communities (grasstops), who could help to build support for the industry. Across the US various groups, such as the Citizens for Responsible Energy Solutions or the Texas-based 'Wind Coalition', came out in support of the wind industry (Waller, 2014; Copley, 2015a).

> Until elected officials know that their constituents care about it [the PTC], they're going to probably continue to vote the way that they always have in the past.
>
> (Copley, 2015a, quoting James Dozier, Executive Director,
> Citizens for Responsible Energy Solutions)

However, divisions within the industry remained and some firms, namely GE and NextEra Energy, attempted to hedge their position by advocating for the PTC to be phased-out rather than for a multi-year extension (Anon., 2015). Other firms claimed that those pushing for a phase-out 'buckled too early in the process', because they 'believed it was the only thing you could get out of an increasingly Republican Congress' (Interview 54). As a result, firms such as Vesta and Iberdrola Renewables mobilised their own ad hoc coalition, the aptly named 'ad hoc group', to lobby Congress to extend tax credits for wind and solar (Capitol Tax Partners, 2015). As I will discuss in the next section, this split may have also reflected varying commercial interests between these firms.

With legislative approval required to extend the PTC, the focus remained on the Congress. As Jim Reilly, senior vice president of legislative affairs at the AWEA explained:

> We knew that going into this fall that the House was going to be a challenge, and we were not going to see them initiating a PTC extension. So the bulkhead is the Senate.
>
> (Copley, 2015b)

The difficulty in the House was on display in April 2015 when Republicans Kenny Marchant and Mike Pompeo introduced the *PTC Elimination Act* (United States Congress, 2015). They were supported by conservative groups, such as the American Energy Alliance, which framed the PTC as 'corporate welfare' claiming it was a 'massive handout for large corporations' (AEA, 2015b). Supporters of the industry retaliated framing the issue as one of fairness. For example, Senator Grassley argued that 'opponents of the renewable energy provisions ... disregard the many incentives and subsidies that exist for other sources of energy' (Ward, 2014). Once again, the wind industry had success in the Senate Finance Committee, which in July voted 23–3 to approve a package of tax extensions that included a two-year extension of the PTC (Federal Information & News Dispatch, Inc., 2015: 2).

As these negotiations unfolded in Congress, business actors continued to lobby. For example, in September 2015 wind companies from both camps, such as GE, which supported a phase-out and Iberdrola Renewables, which supported a longer extension, signed a letter organised by NAM from over 2,000 companies calling on Congress to ensure a multi-year or permanent extension of expiring tax provisions, including for renewable tax credits (NAM, 2015). Opponents led by Americans for Prosperity and the American Energy Alliance countered with their own letter urging Congress to end the tax extensions, again using the frame of corporate welfare (Feldscher, 2015). Such efforts on both sides were designed to build the legitimacy of the campaign by highlighting to policymakers that there is a large set of interests in support of their position.

However, by mid-2015 the wind industry campaign was beginning to be affected by the parallel effort of the oil industry to lift the ban on the export of crude oil, discussed in Chapter 3. Firms in the wind and solar industries, with the support of Democrats, saw an opportunity to extend renewable energy tax credits in exchange for permitting the export of crude oil.[5] It was an uneasy trade-off for Democrats with Democrat Minority Leader, Nancy Pelosi, rationalising that the extension of tax credits for renewable energy 'will eliminate around 10 times more carbon pollution than the exports of oil will add' (Lacey, 2015). Republicans came on board as well, recognising that 'even though they wanted to kill' the PTC, a phase-out 'was likely the best they could do', (Interview 52). On 18 December 2015 Congress agreed to the Omnibus Appropriations bill, which established a phase-out of the PTC, and extended the ITC for solar energy. While it was not the multi-year extension that some actors in the wind industry had lobbied for, it was widely lauded as a victory by the industry (AWEA, 2015b).

Why and how did business actors shape the contest over the PTC?

Preferences

Across the industry wind manufacturers and wind power operators supported the PTC because it was in their commercial interests to do so. The PTC reduces the price of wind-generated electricity making it more attractive to investors and more competitive with other forms of electricity, which is why, as one industry representative put it, 'if you're a wind company you support it' (Interview 52). As a result, there was no evidence of significant intra-industry conflict within the industry. However, support for the PTC did not manifest into a uniform position. Specifically, in the context of a Republican controlled Congress, there was a split between those firms that preferred a phase-out of the PTC, and those that preferred a multi-year extension. To some extent, this likely reflected varying commercial interests. The evidence indicates that the principal firms that preferred a phase-out of the PTC were NextEra Energy and GE. In the case of NextEra Energy, some respondents suggested that this may have reflected relative gains (Interview 55). In other words, while NextEra Energy stood to lose from the expiration of the PTC, as the largest wind project asset owner in the country, it may have been in a better position relative to some of its smaller competitors without the PTC. In the case of GE, it may have reflected a strategic choice to prioritise other reforms to the tax code ahead of the extension of the PTC, which it was more likely to achieve by supporting Republican calls for a phase-out (Interviews 55 and 56). In other words, it was an attempt to hedge its position. With the advent of a Republican controlled Congress in 2010 and mounting pressure from Republicans to eliminate renewable energy tax credits, including the PTC, some firms in the industry believed that the best way forward was to lobby for the PTC to be phased-out because a multi-year extension was not possible. While some respondents claimed this was 'mistake' (Interview 55), for several of the largest firms, such as GE, it was an important strategy. Nevertheless, these differences in the industry were minor and did not reflect the type of intra-industry divisions that were present in the policy contests discussed in the previous chapters.

Instead, the principal divisions were inter-industry, particularly with those firms in the electric utility industry that had significant nuclear power in their portfolio. Although some in the electric utility industry, for example Duke Energy, provided conditional support for the PTC (Interview 53), others were opposed. The most vocal was Exelon, the largest nuclear reactor owner in the US (Trabish, 2014). Again the reason was commercial. In some jurisdictions wind power was eroding the profitability of Exelon's nuclear power plants by putting downward pressure on the wholesale price of electricity. The PTC facilitated this by encouraging wind operators to produce power irrespective of the price, which on occasions meant supplying electricity at negative prices to the detriment of utilities, such as Exelon. These divisions within the electric utility industry meant that EEI, the most prominent industry association, was not active in the policy contest (Interview 53).

Strategies

In the policy contest over the PTC business actors relied on a variety of strategies to shape the outcome – see Table 5.1. The effectiveness of these efforts was invariably determined by the resources actors had at their disposal and their political legitimacy. First, actors on both sides of the contest – namely the wind industry and sections of the utility industry – mobilised coalitions and leveraged wider networks. The AWEA was the command centre for the campaign to extend the PTC working to set the legislative strategy and mobilising wind companies across the country. However, as evident in previous policy contests, firms in the wind industry also used ad hoc coalitions to advance their positions, when they were dissatisfied with how the formal industry association, in this case the AWEA, was representing their preferences. The 'ad hoc group', which included firms such as Iberdrola Renewables and Pattern Energy, played this role to make the case for a multi-year extension of the PTC instead of a phase-out. As one industry representative acknowledged 'every industry association has its divisions' and it was no different for the AWEA (Interview 52).

Nevertheless, as the command centre of the campaign the AWEA helped to mobilise state and non-state actors. For example, the Governors' Wind Energy Coalition came out in support of the PTC at the sub-national level, as did other business actors, such as BICEP, which brought large corporations, including Mars and Starbucks, on board in support of the wind industry. By enrolling these prominent business actors the wind industry was able to enhance the legitimacy of its campaign in the eyes of policymakers, not to mention its resources as these corporations brought with them their own lobbying capacity. Interestingly, interviews across the renewable energy sector indicate that there was very little coordination between the wind industry and the solar industry (Interview 55).

In response to the wind industry campaign there was resistance from some sections of the energy sector, such as Exelon in the electric utility industry, but the relevant industry associations, namely the EEI, stayed out of the fray. Instead, opposition to the PTC was led by a network of conservative advocacy groups, such as the American Energy Alliance and Americans for Prosperity, which had the financial backing of the Koch brothers and support from some in the oil and gas industries, such as the API (Interview 8). While the major business groups did not play an active role in the contest, they were in favour of extending business tax credits generally, which lent legitimacy to the AWEA's campaign (NAM, 2015).

Table 5.1 Summary of the policy contest over the Production Tax Credit

	Wind industry	Incumbent industries (especially sections of utility industry)
General industry preferences	Support	Some opposed
Key coalitions mobilised in the energy sector	AWEA, Ad hoc group	–
Key business coalitions mobilised	–	–
Other important actors in the networks	Governor's Wind Energy Coalition, BICEP	American Energy Alliance Americans for Prosperity
Framing strategy	Framed contest around jobs, growth and fairness	Framed around corporate welfare
Lobbying	Both industries spent millions. For example, in the three years to 2016, the AWEA spent around $3.5 million.	Both industries spent millions. For example, in the three years to 2016, Exelon alone spent around $14.8 million.
Outcome	Production tax credit extended and phased-out	

Second, the wind industry sought to frame the contest around jobs and growth in order to influence the policy contest. For example, in 2015 in an opinion piece in *The Hill*, Mike Garland and Susan Reilly, the chair and past chair of the AWEA, proclaimed that 'killing the PTC' would 'kill thousands of new jobs' and devastate local communities via lost revenue (Garland and Reilly, 2015). Relying on tactics that are ubiquitous across the energy sector, the industry strategically used information to support these assertions, for example, via commissioned reports. However, in an effort to re-frame the contest, opponents of the PTC, such as Thomas Pyle, president of the American Energy Alliance, derided the PTC as 'a textbook case of corporate welfare', which has become 'a massive handout for large corporations' (AEA, 2015b). By using the term welfare, opponents attempted to frame the contest in ways that could be grafted onto the negative connotations attached to the concept of welfare in the community. They also deployed information to cast doubt on the wind industry, for example, releasing reports designed to discredit the job figures released by the AWEA (Scheid, 2013). The wind industry hit back by emphasising the principle of fairness, given the decades of subsidies to the fossil fuel industries, an approach that was to be mirrored by the solar industry in the contest over the ITC. In short, business actors on both sides of the debate employed framing as key tool to shape the ultimate outcome, although no side seemed to achieve the dominant frame.

Third, the coalitions of actors that mobilised in support of extending the PTC engaged in both inside and outside lobbying. The AWEA led the inside lobbying effort. It set the legislative strategy, which was to focus on the Senate, especially the Senate Finance Committee, where they had the longstanding support of Senator Grassley, among others. The aim was to ensure that the PTC was 'a Democrat priority' and that the 'Republicans are divided' on the issue, which invariably they were (Interview 52). While the AWEA did not have the financial resources of other industry associations in the energy sector, such as the EEI or API, the largest wind firms were able to marshal significant resources. For example, NextEra Energy, the largest wind developer in the US, spent over $11 million on lobbying in the three years to 2016, though, of course, not all of it was directed at the PTC (CRP, 2017). This was complemented by an outside lobbying campaign, which involved television and radio advertisements (Interview 55). And, as several industry representatives pointed out, they used their wider networks to 'cultivate opinion leaders in local communities' as part of the effort to build not only grassroots support, but support among the so-called 'grasstops' as well, which enhanced the legitimacy of the campaign (Interviews 54 and 52).

Finally, what impact did business actors have on the policy contest? As discussed in Chapter 2, determining business influence is a tricky task. To some extent, the question can be answered by examining business preferences and strategies – in other words, their actions. For example, the agreement reached in December 2015 to phase out the PTC was consistent with the preferences of some sections of the wind industry, and it would not have been achieved without the wind industry mobilising in support of the PTC. Yet, ultimately, the deal on the PTC was facilitated by the campaign to lift the ban on crude oil exports, which Republican legislators were eager to support, and like their Democratic counterparts, were willing to make trade-offs on.

Accordingly, the influence of business actors was conditioned by the mobilisation of other non-state and state actors. First, the wind industry's capacity to shape the contest and have the PTC extended was enhanced by the support of non-state actors including environmental NGOs and citizen groups. As the CEO of the AWEA made clear, the wind industry sought to enhance its influence by cooperating and or seeking the support of outside groups, such as BICEP, which was founded by the environmental NGO CERES. These groups lent the industry's campaign political legitimacy by demonstrating the breadth of support, but also resources with BICEP helping to enlist firms such as Mars and Starbucks in support of the campaign.

Second, policymakers, too, created and limited the opportunities for business actors to shape outcomes. For example, the decision by some firms, such as NextEra Energy and GE, to support a phase-out of the PTC, and the associated positioning of the AWEA, reflected the view that in the context of a Republican controlled Congress this was the best that could be hoped for. While the White House and most Democrats in Congress supported an extension of the PTC, especially those that had state-based incentives to do so by virtue of the industry's presence in their jurisdiction, many Republicans did not. This meant that a multi-year extension of the PTC would be harder to achieve, though not impossible.

The policy contest over the ITC: 'it's gonna be a slugfest'[6]

Much like the wind industry, the solar industry has also benefited from tax credits. The most important has been the ITC, which reduces federal income taxes by 30 per cent for capital investments in solar systems on residential and commercial properties. Established in 2006 under the Energy Policy Act of 2005, it was extended by President Bush in the midst of the global financial crisis for eight years to 2016 (SEIA, 2015c).[7] The ITC has been a boon for the industry, and it is no surprise that as the expiration date drew near the solar industry fought to have it extended. Paralleling the contest over the PTC, the White House supported a further extension of the ITC, but a Republican majority in Congress did not. The contest pitted the emerging solar industry against the incumbent industries, specifically the utility industry. While many solar firms had given up hope that the contest could be won in early 2015, by the end of the year they were celebrating an unlikely victory as Congress agreed to extend the ITC to 2022, as part of the Omnibus Appropriations bill, which extended the PTC.

The campaign to extend the ITC began in 2013. SEIA, which all respondents described as the coalition leading the campaign, mapped out the strategy (Interviews 1, 40, 44, and 45). The aim was to mobilise the 1,000-plus firms that were members of SEIA and its state-based chapters to build support for solar among Democrats and Republicans in Congress (Lacey, 2016). In October 2014, the mobilising began. In a widely reported speech, Rhone Resch, President of SEIA, launched a call to arms:

> It's absolutely imperative ... job #1 ... that we extend the 30 per cent solar Investment Tax Credit past 2016. Let me be blunt: It's not going to happen without your help. We need each and every one of you – now ... and more than ever. So, today is the official start of our round-the-clock campaign to 'Extend the ITC'.
>
> (Resch, 2014)

To make their case, SEIA initially framed the contest around fairness (Interview 25).

> 100 years ago ... Our friends in the oil and gas industry began capitalizing on a huge tax break.... Well, guess what? 100 years, and hundreds and hundreds of billion dollars later, big oil is still reaping this very generous benefit – along with many others buried in the U.S. Tax Code. Explain to me: How is that fair? 100 years for big oil. But just 10 years for the 30 per cent solar ITC.
>
> (Resch, 2014)

However, the solar industry was fighting an uphill battle. The main opponent was the utility industry, which viewed the expansion of solar as a threat. As SEIA began to mobilise to extend the ITC, so, too, did the EEI, but their aim was to end it and they had considerable financial resources at their disposal. For example, in the three years to 2016, the EEI marshalled around $27 million for lobbying compared to about $2.5 million by SEIA (CRP, 2016d, 2016i). The frustration among solar lobbyists was evident, with one despairing that 'EEI is gigantic ... we have 10 people versus 100 people in EEI' (Interview 41). Even among those utilities that benefited from the ITC the decision was made to stay on the side-lines and leave the lobbying to the solar industry (Interviews 19 and 40).

Further, as the CEO of one major business associations pointed out 'all mainstream business groups are lined up on energy policy. The outliers are wind and solar associations who want to preserve subsidies' (Interview 11). Across the oil, gas, coal, and utility industries executives argued that it was time for renewables to 'get off the dole' (Interview 26) and 'stand on their own' (Interview 21) because in their view solar is already competitive

(Interview 8). Following its 'end wind welfare campaign', the American Energy Alliance, which as discussed has ties to the fossil fuel industries, indicated that it too was working to end the ITC (SEIA, 2015a; AEA, 2015a).

In February 2015, President Obama outlined his budget for 2016, which included a permanent extension of the ITC (Office of Management and Budget, 2015). The announcement was hailed by SEIA as a 'major investment in America's future' (Carvill, 2015). In May, Democrats in Congress introduced a bill extending the ITC for five years. Yet there was little hope that either would be passed by the Republican controlled Congress (Lacey, 2016). According to solar lobbyists, a 'decision was taken to reach out to Republicans' (Interview 1). Again, SEIA led, hiring several Republican aligned lobbyists from law firm Squire Patton Boggs, including Trent Lott, the former Republican Senate Majority Leader (Lacey, 2016). It also used its SuperPAC – SOLARPAC – to target Republicans in Congress, such as Republican Senate Finance Chairman, Orrin Hatch, who had taken over as Finance Chairman from Democrat Senator Wyden in January 2015, and was a strong supporter of solar power (SEIA, 2015a). Since 2012, the SOLARPAC spent over $200,000 targeting Democrats and Republicans in Congress (CRP, 2016h).

Building on the strong public support for solar power, with eight out of every ten Americans wanting more emphasis on solar power, the industry worked to build a grassroots campaign that mobilised Republican voters (SEIA, 2015b). This was made easier by the fact that a majority of Republican voters supported solar power and grassroots groups had already emerged to advocate for solar power in Congress. In 2013, former Republican congressman Barry Goldwater Jr. became chairman of Tell Utilities Solar Won't Be Killed (TUSK), which has fought utility efforts to stifle solar. The Atlanta Tea Party also teamed up with the Sierra Club, one the largest environmental NGOs in the US, to establish the Green Tea Coalition, which advocates for renewable energy (Gilbert, 2014).

The solar industry's framing also shifted to focus on jobs and growth. Solar firms argued that not only is the solar industry 'literally contributing to the job growth of the U.S. economy' (Korosec, 2015), but that thousands of jobs would be lost if the ITC was allowed to expire (Interview 44). Mirroring what we saw in previous chapters, the rhetoric was backed up with reports detailing the employment contribution of the industry. For example, the Solar Foundation, an industry advocacy group, released its annual solar census update for 2014, which claimed that the industry is 'adding workers at a rate nearly 20 times faster than the overall economy' and now employs over 170,000 Americans, contributing some $15 billion to the US economy (The Solar Foundation, 2015).

Despite the campaign, by early 2015 many solar firms believed the ITC's days were numbered largely because Republicans in Congress were determined to eliminate renewable tax credits. As a result, there had been a rush in solar construction as developers tried to capture the 30 per cent ITC before it expired (BNEF, 2015). The industry also attempted to hedge its position by inserting a clause in the existing law to permit projects that 'commence' before the end of 2016 to receive the 30 per cent tax credit even if they did not generate power until after 2016, and to phase down the ITC over several years rather than reaching a cliff in 2016 and plunging from 30 per cent to zero (Cardwell, 2015; Interviews 25 and 44; McTague, 2014).[8] At the same time, firms such as SolarCity sought to raise tax equity funding before the deadline. For example, in 2015, Google invested $300 million in a tax equity fund for SolarCity to install PV panels on residential rooftops (Wang, 2015; Interview 45).[9]

However, by late 2015, as discussed above, the push by the oil and gas industry to lift the restrictions on crude oil exports was gaining strength and the solar industry, like its counterparts in the wind industry, saw an opening. In November, SEIA organised 15 executives from leading solar companies to travel to Washington DC to directly lobby Congress. Leading Democrat Senators, including those who supported oil exports, such as Senator Heidi Heitkamp, proposed linking renewable energy tax credits to the oil export ban (Lacey, 2016). The key was Republicans – as one renewable lobbyist put it, 'you had to get them on board' (Interview 1). And, on 18 December 2015, they came on board to extend the ITC via the Omnibus Appropriations bill. While it may never have happened without the strength of the oil and gas industry, the solar industry had achieved almost everything it wanted; the ITC was extended for six years until 2022; a commence and construct clause was included; and the ITC would gradually decline to 10 per cent in 2022 rather than plunging to zero. The effect was immediate with the share price of leading solar firms soaring (Gross, 2015).

Why and how did business actors shape the contest over the ITC?

Preferences

The solar industry, with the principal exception of First Solar, supported the extension of the ITC for the simple reason that it was in their commercial interests. The ITC lowers the cost of solar power making it more competitive with oil, gas, and coal. Business actors in both the installation segment of the market and the utility-scale manufacturing segment supported the extension. For example, SolarCity, Sunpower, and Sunrun, three of the largest solar installers, all released statements 'urging Congress to act quickly to enact the measure' (SolarCity, 2015; Werner, 2015). So, too, did solar manufacturers, such as SolarWorld, one of the largest manufacturers in the US, which proclaimed the passing of the legislation as 'a big step forward for solar power in America' (SolarWorld, 2015). As a result, there were few signs of intra-industry conflict, because 'on the ITC everyone is pretty well aligned' (Interview 25). Indeed, all the major industry associations such as SEIA and ACORE campaigned for the extension (McGuiness, 2015). At the same time, the solar industry also hedged its position in a similar fashion to the wind industry. When the contest commenced, the solar industry believed that the ITC was likely to expire. As a result, it advocated for the ITC to gradually expire, rather than falling from 30 per cent to zero in 2016, providing more time for the industry to adjust (Interviews 25 and 44). In the end this was unnecessary as the push by the oil and gas industry to lift the ban on crude oil exports provided new opportunities for horse-trading in Congress and the extension of the ITC until 2022.

The principal exception was First Solar, which 'did not favour an extension of the ITC' (Hughes, 2016). There are two likely explanations for its decisions. First, there is some evidence to suggest that prior to the policy contest commencing in 2013 First Solar had decided to shift away from investments that are reliant on government subsidies (Sweet and Chernova, 2011).[10] As a result, First Solar likely had less to lose from the expiration of the ITC than other solar corporations. Second, First Solar had built important relationships with utilities that opposed subsidies for renewable energy and it had benefited from the support of these companies for its utility scale operations, as evident in Arizona (Trabish, 2013). In other words, it operated in an environment that was more closely tied to the utility industry. To be sure, its CEO, James Hughes, had also spent much of his career working in the utility industry.

However, in following their commercial interests and supporting the extension of the ITC, these actors naturally became entangled in inter-industry conflicts, notably with the electric utility industry, which actively opposed the ITC. As noted above, electric utilities led by the EEI opposed subsidies for solar, including the ITC, because solar power threatened their existing business model. In other words, it threatened their commercial interests. Further, they were not alone with other groups associated with the incumbent fossil fuel industries, such as the American Energy Alliance, opposing the extension of the ITC.

Strategies

In order to realise their preferences, business actors in the solar industry mobilised coalitions to shape the policy contest – see Table 5.2. The effectiveness of these strategies was a function of the resources solar firms had at their disposal and their political legitimacy. SEIA, as the principal solar industry association, acted as the command centre, and leveraged its state-based chapters to mobilise hundreds of solar firms across the country to build support for extending the ITC. As its CEO, Rhone Resch, told members back in 2014, it 'is going to be a long, hard, uphill battle, but by

Table 5.2 Summary of the policy contest over the Investment Tax Credit

	Solar industry	Incumbent industries (especially sections of utility industry)
General industry preferences	Support	Opposed
Key coalitions mobilised in the energy sector	SEIA	EEI
Key business coalitions mobilised	High technology firms, e.g. via Advanced Energy Economy	–
Other important actors in the networks	TUSK, Green Tea Coalition	American Energy Alliance
Framing strategy	Framed contest around jobs, growth and fairness	Framed around corporate welfare
Lobbying	Both industries spent millions. For example, in the three years to 2016, SEIA spent around $2.5 million.	Both industries spent millions. For example, in the three years to 2016, the EEI spent around $27 million.
Outcome	Investment tax credit extended and phased-out	

sticking together – and working together – we can be successful' (Resch, 2014). To be successful, and build support among Democrats and Republicans in Congress, the solar industry needed to mobilise other business actors in support of solar power. However, unlike the oil, gas, coal, and utility industries, it did not have the same structural power and could not rely on the most powerful business associations, such as the COC for support because these associations viewed solar as an 'outlier' and sided with the incumbent industries when it came to solar subsidies (Institute for 21st Century Energy, 2015; Interview 11).

Accordingly, the solar industry sought to leverage rest of the clean energy sector. One way was to build and strengthen ties with high technology companies, such as Google and Apple, which are more supportive of renewable energy 'because they invest in it and have skin in the game' (Interview 32). The benefit of establishing ties with these companies is that they give the solar industry what it does not yet possess as a developing industry: political legitimacy. In the words of one solar executive 'if Google can get behind this technology then others can too' (Interview 45). It is no surprise then that clean energy associations, such as AEE, which supported the extension of the ITC, have worked to leverage broader clean energy networks.

> The idea is if you are going to be a player in the debate you have to demonstrate heft. Each sector of the [solar] industry is small compared to the oil, gas, or coal industry, but if you add them all together and we stand together then we are bigger than the pharmaceutical industry, or the movie industry, or the airline industry.
>
> (Interview 3)

As the leader of the campaign, SEIA attempted to frame the contest in order to set the agenda. It identified the expiration of the ITC as the problem and its extension as the solution and it strategically tried to frame the debate around fairness. As the president of SEIA argued: 'we are going to pound away at the fairness argument' (Resch, 2014). And they did. In public commentary and in interviews, industry representatives tied their case for the extension of the ITC to the principle of fairness arguing that they 'want a level playing field' with the oil, gas, and coal industries (Interview 25). Relying on industry reports, solar firms argued that not only is the extension of the ITC fair, but it will contribute growth and jobs to the US economy.

However, this frame did not go unchallenged. As other researchers have pointed out, one of the most effective ways for business actors to respond is to strategically re-frame the contest (Sell and Prakash, 2004). In an effort to do so, firms in the incumbent fossil fuel industries framed the ITC as welfare for the solar industry, just as they had framed the PTC as welfare for the wind industry. As one industry executive argued, renewables have 'alienated a whole group of people and never got off the dole' (Interview 36). Again, the solar industry countered these claims in part by seeking to frame the contest around jobs and growth, though it is unclear if either industry was successful in framing the contest entirely on their terms.

Instead, lobbying was arguably a more important strategy. SEIA led the lobbying effort by mobilising funds and coordinating firms. As noted, in the three years to 2016, SEIA spent $2.5 million on lobbying. These funds were spent on inside lobbying targeting Republicans because, as almost all respondents noted, the problem was not the administration, but the Congress (Interviews 39, 40, and 44). For example, in November 2015 SEIA organised 15 executives from leading solar companies to travel to Washington DC to lobby Congress. This was matched by many of the largest solar companies, such as Solar City, which steadily increased its spending from tens of thousands in the 2012 election cycle to hundreds of thousands in the 2014 election cycle (CRP, 2016f, 2016j). Interviews suggest that a large portion of these funds, especially those from SEIA, were targeted at lobbying on the ITC (Interview 44). However, the resources of the utility industry, which resisted the ITC, dwarfed those at the disposal of the solar industry. As noted, in the three years to 2016, the EEI spent around $27 million on lobbying.

Finally, what impact did business actors have on the contest over the ITC? As was the case with the PTC, part of the answer comes from examining business preferences and strategies. Once again, the agreement to extend the ITC in 2015 was consistent with the preferences of business actors in the solar industry, with the exception of First Solar, and the evidence indicates that the active engagement of the industry in the contest, via framing and lobbying in particular, was vital to the agreement's success. Yet, paralleling what occurred in the contest over the PTC, the success of the solar industry's campaign unwittingly appears to have hinged on the oil industry's efforts to lift the restrictions on oil exports, which Republicans were willing to support even if it meant extending renewable tax credits.

Further, the mobilisation of other non-state actors and policymakers created and limited the opportunities for business actors in the solar industry to shape the contest. First, there is some evidence to suggest that the solar industry cooperated with the other non-state actors, especially in their efforts to combat the resistance from the utility industry, such as the well-named coalition Tell Utilities Solar Won't Be Killed (TUSK). Though for the most part, the role of other non-state actors appears limited, at least in the federal contest over the ITC. Whereas the role of policymakers clearly played a more direct role in conditioning the opportunities for the solar industry to influence the contest. As noted, the White House and Democrats in Congress supported the ITC, as they did the PTC, and they had strong political incentives to do so with polling showing strong support for solar power (Ansolabehere and Konisky, 2014: Ch. 3).

However, the solar industry also faced resistance from Republicans in Congress, which is why they mobilised the resources to target them, particularly in their lobbying approach. These efforts were assisted by leveraging networks with links to the Republican Party, such the Green Tea Coalition, which supported solar power.

Conclusion

The rise of renewable power has set off a series of policy contests in the energy sector, especially between incumbent industries, such as the utility industry, whose profits are threatened by wind and solar. The contests over the PTC and the ITC have been two of the most critical because they have facilitated the development of both industries. Business actors in the wind and solar industries have led the campaign to defend these policies in the face of resistance from other sections of the energy sector. The contests are worthy of attention, not only because they provide an opportunity to examine the behaviour of actors in industries that will be vital to a clean energy transition, but also because they offer insights for how to build green coalitions that might overcome the political resistance of incumbent fossil fuel industries.

Tables 5.1 and 5.2 summarise business behaviour in these policy contests. In the wind and solar industries the vast majority of firms supported the extension of the PTC and the ITC because it was in their commercial interests to do so. In pursuing their interests, inter-industry conflicts broke out with the utilities because the surge in renewable power directly threatened their business model, which in turn led to 'a well-organised backlash' from the utility industry (Interview 44). Despite the split in the wind industry over whether to phase-out the PTC, there was uniform support for the policy, unlike in the solar industry where First Solar was an outlier. On the one hand, this likely reflected that First Solar had less to lose commercially from the abolition of these policies, and on the other, First Solar operated in an environment closely tied to the utility industry. Further, in both contests segments of the industries looked to hedge their position. In the wind industry, the decision by some firms to lobby for a phase-out of the PTC, rather than a multi-year was a classic example given the antagonistic political environment in Congress. For many of the same reasons, there was a parallel effort by the solar industry to shape the rules so that the ITC would gradually expire, providing more time for the industry to adjust, should the policy not be extended.

In pursuing their interests wind and solar corporations mobilised coalitions and leveraged a diverse network of actors in support of their campaigns. AWEA acted as the key coalition in the wind industry and SEIA in the solar industry, with both mobilising hundreds of firms across the country to build support for wind and solar among Democrats and Republicans in Congress. In both cases these actors sought to leverage ties with other actors in the business community, such as investment firms and high technology firms, which helped to increase their resources and enhance their legitimacy as they made their case to policymakers. In part this was driven by the fact that they could not rely on the most powerful business associations for support, such as the COC or the Business Roundtable, because these associations viewed renewable energy as an 'outlier' and largely sided with the incumbent industries.

These coalitions of actors in turn attempted to frame the contest drawing on their discursive power. In the case of the PTC and the ITC, coalitions in the wind and solar industries worked to frame the contests around jobs and growth. For example, the AWEA and SEIA strategically deployed reports to highlight their industries contribution to the economy. In addition, they also relied on fairness as a principle to advance their case pointing to the many years that other energy sources, namely oil, gas, and coal, had received subsidies. This was especially useful when countering claims from the incumbent industries that renewable tax credits amounted to corporate welfare.

As emerging industries, wind and solar did not have the same financial resources as the fossil fuel industries. For instance, the revenues of the largest solar corporations, such as First Solar were in the billions, rather than the hundreds of billions in the oil and gas industry. Nevertheless, in both contests wind and solar firms spent large sums lobbying. For example, as the key coalitions coordinating the lobbying effort AWEA and SEIA spent $3.5 million and $2.5 million respectively in the three years to 2016. Both campaigns targeted Republicans because overcoming their opposition would be crucial to any legislative action on tax credits. For example, AWEA reached out to Republican allies on the Senate Finance Committee. Likewise, SEIA developed a network of lobbyists, which included hiring Republican aligned lobbyists and arranging for solar executives to travel to Washington DC to lobby Congress (Lacey, 2016).

Finally, there is substantial evidence that business actors shaped the policy contest over the PTC and the ITC. Without the campaign by both industries in support of renewable tax credits, it is possible that they would never have been extended. At the same time, without the resistance of some sections of the energy sector, such as electric utilities, the debate over the PTC and the ITC would have been less controversial and a multi-year extension more likely. That said, the deal to extend both tax credits in the end was made possible by the campaign to lift the ban on crude oil exports, which enabled Republican and Democrats to agree to a grand bargain.

However, the influence of wind and solar firms was mediated by the mobilisation of other non-state actors, and the role played by policymakers. While it is not possible to definitively quantify the influence of business actors on the contests, consideration of the context in which business actors operated allows for a more accurate assessment. To varying degrees, the wind industry and the solar industry sought to cooperate with, or were supported by, other non-state actors, be they associated with environmental NGOs in the case of the BICEP or the Green Tea Coalition, or more direct citizen groups in the case of TUSK. These groups appear likely to have strengthened the resources and legitimacy of the renewable energy industry as they lobbied to have the PTC and ITC extended. Equally, other non-state actors associated with the incumbent fossil fuel industries, such as American Energy Alliance or Americans for Prosperity, likely enhanced the campaign of the utility industry in resisting renewable subsidies.

In part, these actors did so by influencing the role of policymakers. While policymakers that supported renewable tax credits often did so by expressing a desire to support clean energy and act on climate change, they also had strong political incentives to do so. For example, most national polls and state polls showed significant support from Republican and Democrat voters not only for renewable energy, but for renewable tax credits as well (AWEA, n.d.; Solar City, 2015; SEIA, 2015b; Ansolabehere and Konisky, 2014: Ch. 3). This afforded opportunities for firms in the renewables sector to mobilise these broad coalitions in support of tax credits and to target their lobbying efforts accordingly. However, their capacity to shape the contest was constrained by the fact that many Republicans remained opposed, even in states that had strong support for wind and solar power. For fossil fuel firms and their networks of conservative advocacy groups, this provided alternative opportunities to reinforce Republican opposition, for example via campaign donations. As several respondents claimed, 'polling is not sufficient to adjust behaviour', and what political leaders 'care much more about is … campaign contributions' (Interviews 52 and 54).

In summary, there are many similarities between the behaviour of firms in the solar and wind industries and the incumbent fossil fuel industries examined in the previous chapters. However, what is striking about these two policy contests is that they reveal not only the preferences and strategies of wind and solar firms, but also the nature of the resistance to renewable energy. Overcoming resistance from incumbent industries and building coalitions to support clean energy technology will be vital if an energy transformation is to occur – a topic I will turn to in the final chapter.

Notes

1 Interview 45.
2 Interview 52.
3 Members of the ad hoc group include Apex Clean Energy Holdings, LLC; EDF Renewable Energy; EDP Renewables NA; E.ON North America; Iberdrola Renewables; Invenergy LLC; Pattern Energy Group LP; Terra-Gen Power; and Vestas-America Wind Technology, Inc. (Capitol Tax Partners, 15 April 2015).

4 'What we see in the Senate Finance Committee markup this week is an opportunity to have [the PTC] extended through 2015 with the start-construction language that would very much help avoid the boom-bust that we've had in the last year or so,' Kiernan said (Copley, 2014).

5 This was not the first time renewable tax credits had been linked to trade-offs with the oil industry. For example, in January 2015 Senator Heitkamp linked approval of the Keystone XL pipeline to a five-year extension of the PTC (Heitkamp, 2015).

6 See SEIA, 2015a.

7 To earn the 30 per cent tax credit the solar system had to commence operation prior to 31 December 2016, when it is due to expire. Under the previous rules, the ITC would then decrease to 10 per cent for commercial properties and to zero for residential properties.

8 This is to provide certainty to large utility scale projects that can take several years to construct.

9 SolarCity has also raised tax equity funding from the finance sector via Goldman Sachs, US Bancorp, and Credit Suisse, among others (Woody, 2013).

10 As the CEO of First Solar James Hughes stated in 2012 when describing the company's strategy: 'let's go to those places where we are ... economic today with no subsidies' (Parkinson, 2012).

6 Re-thinking business behaviour in the US energy sector

Introduction

Many studies have demonstrated the influence of business actors across multiple policy domains, including in the field of environmental politics. This is to be expected given the centrality of business actors to economic activities that invariably have significant environmental implications. As I discussed in the opening chapters, scholars have examined the role of business actors in shaping governance outcomes at the national, international, and transnational levels. Yet there is less scholarship on the behaviour of business actors in individual energy-centric industries, namely the oil, gas, coal, utility, and renewable industries. To the extent that they have been considered, more often than not it has been as part of broader business coalitions. This is surprising given that a third of the world's oil reserves, half the world's gas reserves, and almost 90 per cent of global coal reserves must be left in the ground to avoid the worst impacts of climate change (McGlade and Ekins, 2015).

This book has attempted to redress this blind spot in the literature by examining the behaviour of business actors across these energy industries. Using contemporary policy contests that have taken place during the period of the Obama administration (2008–2016), I have attempted to trace not only why business actors behave the way they do, but also how. In other words, to understand their preferences and to unpack the complex mechanisms by which they seek to influence governance outcomes. The aim has not been to explain policy outcomes per se, which would naturally require the focus to shift to a wider array of state and non-state actors, rather it has been to identify business preferences and strategies and approximate their likely influence. However, business actors do not operate in a vacuum and to the extent that their opportunities to shape outcomes is conditioned by the mobilisation of other non-state actors and the role of policymakers, these interactions have also been explored. In this chapter, the first of two concluding chapters, the aim is to draw together this empirical work to synthesise the theory and evidence. The final chapter will draw out the lessons for policymakers seeking to regulate these industries and advance a clean energy transition.

Why are business actors shaping energy contests in the US?

The empirical evidence presented in this book indicates that the answer is commercial interests. Business actors in the US energy sector seek to shape governance outcomes that improve their competitive position in the markets in which they operate. Of course, the evidence indicates much more. It speaks to existing theoretical expectations that variations in the distributive effects of policies on business actors will lead to divergent preferences and often industry conflict. And, it speaks to how the unique history and culture of a firm or industry can affect how businesses determine their policy positions. Importantly, it also suggests new insights about how business actors determine not simply whether to support or oppose a policy, but also why in many cases they seek to hedge their position. In what follows, I will explore these issues by drawing together the evidence from the case studies.

First, across the oil, gas, coal, utility, wind, and solar industries in the US, there is widespread evidence that more than any other factor, business preferences are determined by commercial interests. In the policy contests over oil and gas exports, producers, such as ExxonMobil, Shell, and Chevron, uniformly supported easing restrictions on exports for the simple reason that it would allow them to access international markets and higher prices in Asia and Europe. In short, it would increase their profits. Similarly in the contests over the Waxman–Markey bill and later the Clean Power Plan, almost all coal producers opposed both attempts at regulation because of the adverse impact it would have on their competitive position in the energy market. In the case of the Waxman–Markey bill, although the projected cost on coal was arguably modest, all coal producers stood to lose. It was much the same with the Clean Power Plan, which coal corporations viewed as an 'existential threat' to the industry (Interview 35). In the utility industry commercial interests also drove business preferences, although preferences varied according to generation portfolios. In other words, they varied according to the extent to which they relied on coal to produce electricity. Among the largest electric utilities most supported the Waxman–Markey bill and the Clean Power Plan, though in part this reflected the behaviour of hedging described below. And it was the same in the policy contests in the renewable energy sector. Major wind and solar firms supported the extension of the PTC and the ITC because it lowers the cost of wind and solar power making it more competitive with other forms of energy, especially coal.

Second, in following their commercial interests business actors invariably became entangled in inter-industry conflicts. In other words, they engage in business battles. For example, in the contest over gas exports petrochemical manufacturers, led by Dow Chemical, opposed unrestricted gas exports because gas was a feedstock and, in their view, exports would result in higher domestic gas prices, higher costs of production, and therefore lower profits. Similarly, inter-industry conflicts emerged in the case of oil exports for the same reason, namely the impact of lifting export restrictions was different on different industries. In the refining industry, firms, such as Valero Energy and members of the CRUDE coalition, opposed exports because any increase in oil prices could increase their costs of production, reducing their international competitiveness and, in turn, their profits (Klesse, 2013). Conflicts also broke out between wind and solar firms and electric utilities. As detailed in Chapter 5, wind and solar power presents a direct threat to the profits of utilities by undermining their business model. As a result, in both contests electric utilities opposed firms in the renewable energy industries. This was most evident in the solar industry, and it led to direct conflict, or in the words of one solar lobbyist, 'we are seeing a well-organised backlash' from utilities (Interview 44).

Third, however, the empirical evidence also shows that in pursuing their interests business actors in the energy sector do not always form a preference to simply support or oppose a regulatory measure. In some cases they hedge their position. Existing research suggests that business actors tend hedge their position by shaping rules to minimise their compliance costs (Meckling, 2015). The policy contests examined provide greater clarity about how this is done. Two approaches stand out. One was for business actors to lobby to shape the implementation of regulations that they believed they could not successfully oppose. The aim was to minimise the cost of complying with the proposed regulations. This was most evident in the utility industry. In the contests over the Waxman–Markey bill and the Clean Power Plan, utilities that generated more than a third of their electricity from coal and stood to lose from these initiatives, at some points opposed the policies and at others they supported it as they worked to shape the policy detail. For instance, American Electric Power publicly supported both initiatives while remaining associated with actors that were campaigning against them.

Another approach of business actors was to seek to delay the abolition of regulations from which they benefited, but believed they could not successfully sustain with their support. This was most evident in the wind and solar industries. For example, fearing that that they would be unable to stop the expiration of the PTC because of mounting pressure from a Republican controlled Congress, some firms in the wind industry pushed for the PTC to be phased-out because a multi-year extension was not possible. In other words, they hedged their position by seeking to shape the rules. Business actors in the solar industry took an almost identical approach because of the same Republican opposition. Accordingly, they lobbied to shape the rules so that the ITC would gradually expire, providing more time for the industry to adjust rather than experiencing a sudden abolition of the ITC.

Finally, across each of these cases there were also outliers. These can largely be explained by the institutional contexts in which firms are embedded, and importantly how the unique history and culture of a firm or industry shapes how they respond to these contexts. This was the case in the refining industry where two of the largest refiners, Phillips 66 and Marathon Petroleum, did not oppose the push to export oil despite deriving significant revenues from refining. The most likely explanation is the unique history of both corporations. Until recently, both had been part of vertically integrated oil producers – ConocoPhillips and Marathon Oil – and the close historical ties these refiners had to these corporations potentially explains their position not to oppose exports. There were also outliers in the coal and utility industries. For example, in the coal industry Rio Tinto was the only large coal producer to support the Waxman–Markey bill and the Clean Power Plan. The evidence suggests that its preference varied for two main reasons. On the one hand, it was a much more diversified energy corporation given that coal represented less than 10 per cent of its annual revenues, and, on the other, the fact that it was headquartered in the UK and not the US meant that it was embedded in a different institutional context, one that was generally more supportive of action on climate change. There were also outliers in the solar industry. First Solar appears to have been the principal firm opposed to the extension of the ITC. This is likely because First Solar was less reliant on government subsidies, though arguably more importantly, it also reflected that First Solar maintained close ties to the electric utility industry and had benefited from their support. In other words, it was part of an environment that did not support the maintenance of these policies.

How are business actors shaping energy contests in the US?

The empirical evidence shows that business actors relied on a handful of strategies, namely mobilising coalitions of multiple actors at multiple levels, framing and lobbying. And it shows that the effectiveness of these strategies was directly tied to the resources business actors had at their disposal and their political legitimacy. To some extent, this is consistent with what political scientists, business and management, and regulation and governance scholars would expect. Critically however, the fine-grained empirical analysis presented here also highlights important nuances about how business actors seek to shape policy outcomes. In what follows, I will consider each of these strategies with a particular emphasis on the nuances in business behaviour.

Mobilising coalitions

Business actors in the energy sector built and organised coalitions and this was a key strategy in seeking to shape policy contests. However, the evidence from the six cases suggests new insights about how they did so. In particular, the policy contests highlight how traditional industry associations often act as the command centre in the campaign tying together networks of actors; how ad hoc coalitions emerge and are prevalent across the sector; and it highlights the important role that coalitions can provide in building the legitimacy of emerging industries, namely renewable industries.

First, in many of the contests the leading industry associations acted as the command centre of the campaign, or what regulation and governance scholars often refer to as the 'governing node' because they bring together different actors to pool resources, share information, and mediate conflicts to achieve a common purpose (Burris *et al.*, 2005; Hervé, 2014; Kauffman, 2017). For example, in the oil and gas industry, the API, a conventional industry association, played this role in the contests over oil exports. In doing so, it relied on the API's legitimacy within the business community, a source of its structural power, to tie together other business coalitions, including the COC and the NAM, which further enhanced its resources and legitimacy. In turn, these actors built transnational networks, which enlisted foreign governments, such as the Czech Republic, whose ambassador testified before Congress in support of overturning the crude oil ban. Similarly, in the renewable industries the AWEA and SEIA, both traditional industry associations, took on this role. In the contest over the PTC, AWEA acted as the command centre of the campaign, it set the legislative strategy, built coalitions, and leveraged networks to increase resources and enhance the industry's political legitimacy. SEIA did the same in the contest over the ITC, by mobilising hundreds of solar firms and building networks with other firms in the clean technology sector to lobby congress.

Second, the policy contests highlight the prevalence of ad hoc coalitions in the energy sector. That is, informal coalitions that are temporary in nature, established to fight a single policy contest, and typically disband once the policy contest is over (Barley, 2010; Mahoney, 2007). While industry associations are relatively easy to detect, ad hoc coalitions are often overlooked in existing studies of business actors. Yet they were employed in almost every policy contest examined. For example, in the oil and gas sectors Dow Chemical created America's Energy Advantage to represent petrochemical manufacturers opposed to gas exports. Oil refiners established the CRUDE coalition to resist efforts to overturn the export ban. And in response, oil producers created another ad hoc coalition, PACE, to advance the case for exports. A number of respondents concluded that such coalitions are now 'increasingly common' (Interview 35).

The policy battles in the energy sector not only highlight the prevalence of these coalitions, but they also shed light on several interrelated factors behind why business actors decide to establish them. One reason business actors create ad hoc coalitions is because their existing industry associations take a position they do not support. For instance, firms in the wind industry, such as Iberdrola Renewables and Pattern Energy, used the so-called 'ad hoc group' to represent their position on the PTC because they were dissatisfied with the position taken by the formal industry association, AWEA. The same reason underpinned the creation of America's Energy Advantage and the CRUDE coalition in the oil and gas sector. A second reason business actors establish such coalitions is because existing coalitions are not sufficiently engaged on the issue, in some cases because they cannot reach a common position. For example, the creation of America's Energy Advantage by petrochemical manufacturers also reflected the fact that AFPM, which represents firms, such as Dow Chemical, as well as oil and gas producers, did not take an active role in the policy contest because of the diversity of its members' interests (Interview 35). Another factor behind business actors' creation of these coalitions is to act as a command centre for a new campaign. The Partnership for a Better Energy Future is a case in point. It was primarily established by the COC and the NAM as a temporary group to coordinate business opposition to the Clean Power Plan (PBEF, 2015). In doing so, it tied together more than 200 coalitions including state and national associations from the mining, manufacturing, transport, farming, oil, and gas sectors, not to mention dozens of state chambers of commerce. Its power derived from its capacity to coordinate multiple actors together in order to pool resources and enhance the legitimacy of the campaign.

Third, the empirical evidence also highlights the important role that coalitions can provide in building the political legitimacy of emerging industries. The solar industry was a good example. As an industry it was unable to mobilise the most powerful business associations – the COC, the NAM, or the Business Roundtable – in the same way that the oil, gas, and coal industries were because solar remained an 'outlier' in the business community (Interview 11). As a result, it sought to strengthen ties with other coalitions that support clean energy, such as AEE, which included firms in the technology sector, manufacturing sector, and retail sector. Individual solar firms, such as SolarCity, also saw the value in partnering with high technology firms, including Google, because they provided much-needed legitimacy to the solar industry's campaign. The wind industry took the same approach, as discussed in Chapter 5, by mobilising sections of the investment community and large multinationals, such as Mars and Starbucks, to enhance their position in the eyes of policymakers. The greater the political legitimacy of these industries, the greater their structural power, which in turn means they are more likely to have institutional access to state actors (Mügge, 2011: 56–57). While this is particularly important for emerging industries, business actors in the incumbent fossil fuel industries tried similar strategies. For instance, American Electric Power teamed up with trade unions in a bid to build legitimacy with Democrats in the contest over the Waxman–Markey bill (Interview 6).

Framing

A second strategy that was ubiquitous across the energy sector was framing. Consistent with the literature from other policy domains, business actors in the US energy sector regularly engaged in framing contests where coalitions of actors on either side of the contest compete to strategically frame debates to set agendas and draw attention to their concerns. In each of the six cases business actors worked to frame the policy contest on their terms and those opposed sought to counter-frame to replace the existing frame with their preferred frame. As discussed in Chapter 2, frames are important because they help to provide a shared understanding of reality that structure how people behave.

In many of the cases it was not clear that one set of business actors succeeded in achieving the dominant frame. For example, in the contests over the extension of the PTC and ITC, firms in the wind and solar industries were able to frame the contest around jobs and growth to argue that the expiration of these tax credits would negatively impact the economy. However, attempts at counter-framing by conservative advocacy groups claiming that the tax credits were a textbook case of corporate welfare also appear to have gained traction. The exceptions were the battles in the oil and gas sector. In both cases the evidence suggests that oil and gas producers had considerable success in framing the solution to export restrictions as free trade. One clue to their success is that those opposed were often forced to frame their arguments in the context of free trade. For instance, oil refiners that opposed the export of crude oil argued that oil could never be traded freely in a global market controlled by OPEC (Interview 16).

While it is not always possible to determine if one set of actors achieved the dominant frame in a contest, it is possible to identify the tactics they used in their efforts to do so. In the policy contests in the energy sector three stand out. First, business actors employed frames in ways that can be grafted onto existing principles that have normative appeal among actors whose support is useful, such as policymakers. For example, for oil and gas producers free trade was effective because it enabled 'exports' to be grafted onto the established principle of 'free trade', which had widespread normative appeal. This was one of the reasons other coalitions, such as NAM, joined the oil and gas campaign. NAM had supported free trade since its founding in 1895 (Eisenberg, 2013b). Business actors employed the same tactic in the contests over renewable tax credits. Wind and solar coalitions, in part, tied the extension of the PTC and the ITC to the principle of fairness, claiming that such subsidies were only fair given the decades of tax credits to the fossil fuel industries. Conservative advocacy groups and their supporters in the incumbent industries countered by framing government support for wind and solar power as welfare, an existing frame that had negative connotations in the community and among Republican policymakers (Interviews 21 and 26).

Second, what appears to be important in each of the contests is not only the normative superiority of a frame, but also how frames are employed, and ideas disseminated to set agendas (Sell and Prakash, 2004). Across the energy sector business actors strategically deployed information to help reinforce their attempts at framing. Business actors funded think tanks, consultants, and industry groups to provide reports that supported their case. For example, the case for oil exports had to overcome the public's concern that exports would lead to higher gasoline prices. Accordingly, as discussed in Chapter 3, oil producers, their industry associations, ad hoc coalitions, and partners in the energy sector funded a series of economic studies concluding that this would not be the case. Similarly, in the coal industry business coalitions relied on a series of economic studies to inform public debates and disseminate information to policymakers about the potential costs of the Waxman–Markey bill and later the Clean Power Plan. A widely cited study by the Heritage Foundation, for instance, a conservative think tank, estimated that the Waxman–Markey bill would slash real GDP by $9.4 trillion and increase unemployment by almost 2.5 million by 2035 (Beach *et al.*, 2009). While such estimates were vastly different to those of the EPA, they helped to strengthen the strategic frame by contesting the economic rationale for the policy and hence reducing the willingness of legislators in Congress to support it (Layzer, 2012). Examples also abound in the renewable energy industries, with the AWEA and SEIA commissioning studies to highlight the contribution of the wind and solar industries to economic growth and job creation.

Finally, in some cases business actors employed frames to link issues. Oil and gas producers used the frame of free trade to draw attention to the geopolitical advantages that would accrue from providing oil and gas to US allies in Europe and Asia, who were reliant on Russian production. This was reinforced by their efforts to enlist Eastern European governments to their cause, including Hungary, the Czech Republic, the Slovak Republic, and Poland. Likewise, business actors in the coal and utility industries linked the Clean Power Plan to energy security exploiting extreme weather events, such as the polar vortex in 2014, to warn of the risks of targeting coal-fired power plants. And, though it was not central to their campaign, firms in the wind and solar industries worked to link renewable tax credits to climate change campaigns.

Lobbying

Business actors in the US energy sector also lobbied and they did so relentlessly. The empirical evidence shows that firms, coalitions, and the wider networks engaged in the types of inside and outside lobbying that are commonly identified in the political science literature (Baumgartner *et al.*, 2009). For example, business actors in each of the policy contests devoted considerable resources to building personal relationships with policymakers in Washington DC in order to shape the policy outcome – so-called inside lobbying. A classic example is the regular visits that CEOs made to Capitol Hill in every contest to advance their cause. Likewise business actors engaged in outside lobbying, which refers to the public campaigns that business lobbyists organised with advertising, media, and grassroots mobilisation to influence both the public and policymakers. Again, this was common practice among coalitions in all of the cases examined, as evident with energy citizen rallies that oil and gas producers organised, the advertising campaigns of the coal and electric utilities, and the grassroots campaigns run by wind and solar firms.

However, the empirical evidence also shows important nuances in the lobbying behaviour of business actors in the US energy sector. First, is the unequal lobbying capacity of incumbent fossil fuel industries relative to emerging renewable industries, which is a function of the unequal financial resources at their disposal. Take the incumbent fossil fuel industries first. In 2015, three oil and gas corporations operating in the US – Shell, ExxonMobil and BP – made the top ten on the Global Fortune 500 list, and many had revenues in the hundreds of billions of dollars. Shell topped the list with $421 billion (Shell, 2015). This revenue was drawn on to spend extraordinary sums on lobbying. Between 2010 and 2015 the oil and gas industry as a whole spent more than $140 million per year on lobbying (CRP, 2015c). The coal and utility industries have significant financial resources too. For example, in 2015 coal producers had annual revenues of $35 billion and electric utilities $386 billion (Ulama, 2015: 4; Witter, 2015c: 3). In comparison, the emerging wind and solar industries' lobbying capacity was small given their more limited resources. For example, none of the solar corporations made the Global Fortune 500 list and the revenues of the largest solar corporations, such as First Solar were in the billions, rather than the hundreds of billions (Osten, 2015).

Second, and related, there is also an unequal lobbying capacity between fossil fuel industries, largely driven by the structural decline of the coal industry, described in Chapter 4. For example, between 2009 and 2010 during the debate over the Waxman–Markey bill the coal industry as a whole spent $34 million lobbying compared with $18 million between 2014 and 2015, the period of debate over the Clean Power Plan (CRP, 2014). While there are several potential reasons, such as more limited lobbying opportunities for regulations than legislation, which I will turn to below, interviews with coal lobbyists confirmed the view that it was the declining financial resources of the industry that was the cause. As one coal lobbyist argued in 2015 'coal is so strapped for cash at the moment … they are not doing anything federally' (Interview 49). Another referred to the ACCCE as the 'incredibly shrinking coal association because coal is getting clobbered right now' (Interview 10). It had been one of the key coalitions organising the lobbying effort during the Waxman–Markey bill.

Third, the policy contests highlight the critical relationship between the coalition that acts as the governing node or command centre of the campaign and the coordination of lobbying activities. In the oil and gas industries, respondents pointed out that the API and ANGA coordinated much of the lobbying effort, such as the 12 state advertising blitz that the API organised to pressure Congress on oil exports (Interview 8). Similarly, in the coal industry the ACCCE spent almost $15 million working to defeat the Waxman–Markey bill and the Partnership for a Better Energy Future coordinated many of the activities designed to defeat the Clean Power Plan (Interview 10). In the contest over the PTC, the AWEA set the inside lobbying strategy and coordinated much of the outside lobbying via television and radio advertisements and grassroots campaigning. Similarly, in the case of solar, SEIA as the key coalition coordinated the lobbying effort spending $2.5 million in the three years to 2016 (CRP, 2016i). This included leading the inside lobbying activities designed to build Republican support in Congress to extend the ITC. For example, SEIA established a network of lobbyists, hiring Republican aligned lobbyists, such Trent Lott, the former Republican Senate Majority Leader (Lacey, 2016).

Finally, the empirical evidence shows how the structure of the policy contest presents different opportunities to lobby, as the battles in the coal and utility industries demonstrated. The Waxman–Markey bill provided more opportunities to lobby because it was a legislative initiative that required congressional approval, which meant that both coal producers and utilities could target members and constituents to pressure Congress. In contrast, the options were more limited with the Clean Power Plan because it required regulation not legislation. As one executive explained, 'there are more avenues to lobby on a bill, but there is much less on a regulation because you can only lobby the administration' (Interview 14).

Consequently, business actors engaged in forum-shifting, which refers to how actors take actions in different forums to influence governance outcomes (Baumgartner and Jones, 1991; Braithwaite and Drahos, 2000). In other words, when lobbying in one forum failed, business actors simply tried another. One forum was state legislatures. Recalling that state governments are required to implement the Clean Power Plan, the Partnership for a Better Energy Future developed a model bill for state governments that would require state legislatures as well as the state environment agencies to approve any implementation plan. The aim was to open up additional avenues to lobby and veto the Clean Power Plan at the state level. Another forum was the courts given the adversarial nature of the legal system (Kagan, 1991). As one respondent put it 'the threat of litigation is always an option and we keep it in our back pocket' (Interview 7). In the coal and utility industries business actors, especially in the case of the Clean Power Plan, regularly used state and federal courts to challenge the EPA regulations. As discussed, these actors, often in conjunction with state governments, filed more than 100 separate claims challenging the

EPA's authority to regulate greenhouse gas emissions. In most cases they were defeated, but they did succeed in delaying the development of the Clean Power Plan and creating a sense of uncertainty around the regulations, which supported the campaign against them.

Finally, it is worth briefly returning to the interaction between strategies and the means by which they reinforce each other. As discussed in Chapter 2, one of the principal reasons business actors mobilise coalitions is to leverage other strategies given that coalitions can help to pool resources and build legitimacy. For example, coalition building will enhance the effectiveness of lobbying efforts when they increase the financial resources available to an industry, as was evident in the policy contests examined here. Likewise, when coalitions mobilise actors that enhance the legitimacy of their cause, as was manifest when renewable energy industries teamed up with firms in high technology industries, they enhance the effectiveness of framing given that legitimacy is a source of discursive power. Other strategies can also interact to reinforce each other. For instance, framing can help to mobilise coalitions, especially when the frame is tied to an existing principle that has normative appeal. In the policy contest over oil and gas exports, producers framed the contests around free trade, which worked to mobilise other business associations that had historically supported the principle of free trade, such as the NAM.

Assessing the influence of business actors

The preceding sections have considered why and how business actors behave in the US energy sector. This naturally leads to questions about business influence. In other words, to questions about what impact business preferences and business strategies had on the outcome of each policy contest. While assessing business influence is no easy task, the empirical evidence indicates that in each case business did shape the outcome. Indeed, in the oil and gas industries it is hard to imagine restrictions on oil and gas exports being eased without the campaign of producers. It also hard to imagine the Waxman–Markey bill passing in the House without the support of key firms and coalitions in the utility industry, just as it is difficult to explain the ultimate failure of the bill and the Clean Power Plan without taking account of business opposition. And, in the wind and solar industries business influence was also evident. While coalitions of renewable actors may not have been solely responsible for the extension of the PTC and the ITC, the evidence shows they did influence the shape of the ultimate outcome. To be clear, this is not to say that business actors have dictated the outcome in each case; other factors outside the scope of this analysis were also at play, but they have had an impact.

However, aside from concluding that business actors have influenced the policy contests, the empirical evidence also speaks to the role played by other non-state actors and by policymakers in creating and limiting the opportunities for business actors to shape outcomes. In considering the evidence, it is worthwhile reflecting on how business actors can seek to affect those opportunities either by cooperating with other non-state actors or by targeting the beliefs of policymakers and their political incentives. First, while business actors will face competition from other non-state actors that can limit their influence, as coal producers have from environmental NGOs, the mobilisation of non-state actors can also provide opportunities for business to enhance their influence. The policy contests in the energy sector highlight several examples. Consistent with previous research, business actors can build their legitimacy by working with other non-state actors, such as environmental NGOs (Raustiala, 1997; Meckling, 2011). For example, in the contest over emissions trading the position advocated by electric utilities was likely viewed as more legitimate in the eyes of policymakers because it was made by USCAP, a Baptist and bootlegger coalition of environmental NGOs and business actors. Not only does this demonstrate the breadth of support for the position, but it also shows that typically antagonistic interest groups have reconciled their positions (Hula, 1999). The same forms of cooperation can also build resources. This is likely more important for renewable energy industries, whose financial resources to lobby, for instance, are often overwhelmed by the incumbent fossil fuel industries. Hence the solar industry's campaign to extend the ITC was enhanced by the support of various citizen groups, which brought additional resources to bear on the contest. This will likely also be important in intra- and inter-industry battles when smaller sections of an industry are outgunned by large multinationals, as evident in the battle between the large oil producers and their smaller cousins in refining industry.

Second, given that the influence of business actors will also be conditioned by the role of policymakers, which were arguably more significant in the cases examined here, business actors would be wise to target their beliefs and political incentives. The empirical evidence confirms existing research that indicates that the beliefs or ideas that policymakers have about a particular issue or policy will impact their preferences (Layzer, 2012; Sabatier, 1988). For example, in Chapter 4 the support of the Obama administration and Democrats in Congress for the Waxman–Markey bill and the Clean Power Plan was partly driven by their belief in the need to limit greenhouse gas emissions and tackle climate change, which in turn affected the strategies business employed in the contests. Accordingly, to manipulate the role of policymakers and hence the opportunities to shape policy outcomes, one option for business actors, or any actor for that matter, is to target policymakers' beliefs. Among the strategies discussed in this book, framing is likely to be effective because it can help to provide a shared understanding of reality that structures how policymakers behave. This is likely to be especially effective in the early stages of a policy contest when policymakers have limited knowledge about the issue being contested and are yet to form a specific position (Downie, 2012).

Further, given that policymakers will also be driven by their political incentives, business actors can work to shape public opinion as well. Existing research shows that one of the most important factors explaining whether business succeeds or fails in a policy contest is public opinion because policymakers tend to follow what voters want (Smith, 2000). However, this assumes that business is unified and has a single preference, which is manifestly not the case on many issues in the US energy sector. So what should business do? Should they simply seek to shift public opinion to ensure their position has the greatest public support and hope policymakers follow? The answer is not so clear-cut. Previous scholarship indicates that business actors are more likely to determine policy outcomes when a policy issue has low salience with the public. In other words, when it is not on the front pages of newspapers, televisions screens or Facebook feeds. When this is the case, policymakers will weigh more heavily the preferences of organised interest groups rather than the public (Harrison and Sundstrom, 2007; Trumbore, 1998; Culpepper, 2010). Consequently, in the business battles examined in this book, the best approach for those industries with policy preferences that are not supported by the public, or seem unlikely to be supported, is to keep the issue off the agenda so that policymakers attend to their narrow interests. Hence, inside lobbying may be a successful strategy. In contrast, for those industries that prefer a policy that is more popular with the public, such as the solar industry, the better approach could be to try and raise the salience of the issue with outside lobbying, such as the public campaigns that actors organise via advertising, the media and grassroots mobilisation. Of course, once the issue has salience, business actors on both sides will likely devote considerable resources in their attempts to shift public opinion.

Another way to influence the political incentives of policymakers is financial contributions, especially in the US political system. As discussed in Chapter 5, a number of lobbyists and former staffers in Washington argued that policymakers are increasingly concerned about securing the financial resources for campaigning (Interviews 52 and 54). And an increasing body of empirical evidence supports this, with one study estimating that members of Congress spend between 30 and 70 per cent of their time on fundraising activities (Lessig, 2011). However, actors seeking to use financial contributions will need to do so in a targeted way, and to think carefully about what they get in return. For example, is it to buy access? To elect legislators sympathetic to their policy positions and defeat those who are not? Or simply with the hope of favours down the track (Ansolabehere *et al.*, 2003)? In short, while financial contributions may prove a successful strategy in some circumstances, in many cases outside lobbying may be more important.

Conclusion

The policy contests in US energy sector provide an excellent window into business behaviour. The battles that are taking place shed light on how and why firms are behaving in a sector that is vitally important and yet has often been neglected in environmental and energy politics. The aim of this chapter has been to synthesise the theory and the evidence. It is now possible to answer why energy corporations behave the way they do. Business actors follow their commercial interests. Further, in following their commercial interests they invariably became entangled in inter-industry conflicts. In other words, the variations in the impacts of policies on business actors lead to divergent preferences and industry conflict. The empirical evidence also shows that in following their interests firms in some cases hedge their position, particularly when they believe they are unable to prevent the implementation of a regulation, or maintain an existing regulation. In other cases, the institutional context in which business actors are embedded, and how they respond to those contexts as a result of their unique history or their country of origin, can lead firms to develop positions that are unexpected. Put differently, the position of outliers will not always be predicted by commercial interests alone.

It is also possible to answer how business actors seek to shape governance outcomes in the energy sector. That is, to identify the complex means by which firms exercise influence over the policy process. Overwhelmingly, business actors build coalitions to advance their cause and the empirical evidence suggests insights about how they do so. For example, it shows how traditional industry associations often act as the command centre of business campaigns building coalitions of other business actors and leveraging networks at multiple levels, including state actors at the transnational level, to pool resources and build legitimacy. The policy contests also highlight how ad hoc coalitions emerge and are prevalent across the sector, a type of coalition that is often overlooked in existing studies, and it highlights the important role that coalitions can provide in building the legitimacy of emerging industries, namely renewable industries.

In addition to coalition building, business actors also rely on framing and lobbying. Framing is employed by business actors to set policy agendas in ways consistent with their preferences. To do so, they rely on a variety of tactics including using frames that can be grafted onto existing principles that have normative appeal, strategically deploying information to reinforce these frames, and using frames to link issues. Business actors also lobby ceaselessly to persuade policymakers to support their policy position. In the US energy sector, the empirical evidence highlights not only the various forms of inside and outside lobbying, but importantly the unequal lobbying capacity of different industries, which reflects the variations in available financial resources. It also shows the significant role played by the coalition that acts as the command centre of the campaign in coordinating the lobbying effort, and how business actors shift their lobbying efforts to forums where they are likely to have the greatest influence. All of these strategies interact, and they can work to reinforce one another.

A more difficult question to answer is what influence did business actors have on each of the policy contests? Briefly, there is ample evidence to show that business actors in the energy sector did shape the outcome of each of the contests. To be clear, that is not to say they dictated the policy outcomes, but they did influence the end result. However, their opportunities to do so were conditioned by the mobilisation of other non-state actors and the role of policymakers. The cases examined show how this happens, and they also highlight how business actors can cooperate with non-state actors to build their resources and legitimacy, and how they can work to target the beliefs and incentives of policymakers in order to influence the role policymakers play.

Understanding how and why business actors in the US energy sector behave is important not only because the US is an energy superpower and what these actors do often has ripple effects around the globe, but also because it can contribute to renewed theorising about the role of business actors in environmental energy politics. Such a fine-grained empirical analysis also has significant implications for policymakers seeking to regulate these industries; industries that are critical to achieving an energy revolution. In the next chapter I turn to consider the policy implications.

7   What should policymakers do?

Introduction[1]

As I write this final chapter, the need for policymakers to act is as urgent as ever. In the US, fossil fuels continue to represent 80 per cent of primary energy consumption and this has hardly changed in recent decades. Further, given the success of incumbent fossil industries in shaping US energy policy detailed in this book, it is perhaps no surprise that recent data from the US Energy Information Administration projects that carbon dioxide emissions from the energy sector will rise almost 2 per cent in 2018 and remain virtually unchanged the year after (EIA, 2018f). It is a similar story at the global level, with fossil fuels continuing to dominate and global-energy-related emissions rising slightly in 2018 and again in 2019, driven to a large extent by China and India (EIA, 2018f). This state of affairs is likely to be compounded by the election of Donald Trump in 2016 to replace Barack Obama as president. Indeed President Trump's decision to withdraw the US from the Paris Agreement, the most ambitious international climate change agreement in history, in addition to a raft of domestic policies designed to promote fossil fuels, will ensure that achieving an energy transition in the US will remain as hard as ever (The White House, 2017c).

Putting aside the recent upheavals in US politics for the moment, the question for policymakers then is what should they do to advance a clean energy transition, especially when efforts to regulate the energy sector are so fiercely contested? In the US, much like the rest of the world, the key will be to successfully implement policies that encourage the widespread deployment of clean energy, and crudely speaking, reduce the reliance on dirty energy. However, to do so policymakers will have to overcome the resistance of incumbent industries. As previous research on energy transitions has shown, opposition from incumbent industries could delay and even derail attempts at transitioning to a clean energy future (Hess, 2014).

Accordingly, in this chapter the aim to build on the insights of the preceding analysis to identify specific strategies for policymakers seeking to build green coalitions and networks. Specifically, these are to: entrench and build existing interests via targeted sector specific policies; exploit inter-industry and intra-industry divisions with smart policies that, for example, target politically weak industries; and shift existing interests with policies that induce changes in industry investment and structure by sending direct and repeated policy signals. In what follows, I will first consider these strategies and the ways that they can be designed to limit resistance from incumbent industries. Further, in the wake of Donald Trump's election, it is also worth considering the lessons for business actors in emerging renew-able industries, and the strategies they can adopt in the absence of attempts by federal policymakers to advance an energy transition. Finally, the chapter concludes with some reflections on the implications for future research and the implications for a clean energy transition.

Policy strategies for a clean energy transition

A clean energy transition broadly involves a fundamental change in the energy system away from fossil fuels toward the extensive deployment of clean energy. While there is an ongoing discussion around the precise defi-nition of an energy transition, it is widely accepted that it will be difficult, and that time is running out (see, for example, Sovacool, 2016). The diffi-culty of the energy transition results from the 'carbon lock-in' that indus-trialised nations have experienced, which favours fossil fuels and complicates the emergence of new technologies (Unruh, 2000). And the urgency stems from the irreversible damage caused by the growth in green-house gas emissions, which must be limited immediately if we are to avoid the devastation of a much warmer world (IPCC, 2014). Indeed, it is the urgency of the challenge to reduce greenhouse gas emissions that separates an energy transition from most other social challenges.

It is also widely accepted that to achieve such a transition government industrial policy is needed to accelerate the restructuring of industrialised economies toward environmental sustainability (Hess, 2014). In other words, governments must intervene because markets alone have failed to bring about a fundamental change in the energy system. Traditionally policymakers, at least initially, tend to favour climate measures that econo-mists regard as efficient. Governments around the world have favoured policies that place a price on carbon, such as emissions trading or carbon taxes, because they are considered the most efficient way to reduce emis-sions (Stern, 2007). However, while emissions trading may be the most economically efficient policy, it is not always politically successful. Examples abound in North America and Europe of failed attempts to introduce carbon taxes and emissions trading, and more recent cases, such as in Australia, where emissions trading was implemented and then repealed two years later (Crowley, 2017).

One of the principal reasons for the succession of failures has been political resistance from incumbent fossil fuel industries. Scholars of environmental politics and energy transitions have charted the power and influence of business actors in shaping policy outcomes in the domain of energy and climate policy (see, for example, Aklin and Urpelainen, 2013; Falkner, 2008; Hess, 2013; Tvinnereim and Ivarsflaten, 2016; Pegels and Lütkenhorst, 2014). For example, Aklin and Urpelainen (2013) argue that cleaner forms of power production are less likely when fossil fuels dominate the electricity market because of the political power and resistance of incumbent fossil fuel industries. Likewise, Hess (2013) claims that because energy transitions are politically contested, incumbent regimes may resist and alter the trajectory of development. Further, Hughes and Urpelainen (2015) have shown that the 'political-economic clout' of industry, both the fossil fuel industry and the renewable industry, is critical to explaining the variation in energy-related climate policies implemented in industrialised countries.

This reality should cause policymakers to consider measures that build support for more efficient policies. In the case of energy, policy choices that improve the economic competitiveness of clean energy and entrench, create, and expand constituencies, such as renewable industries, who demand support for clean energy and help to drive an energy revolution (Aklin and Urpelainen, 2013; Levin *et al.*, 2012). There is growing empirical evidence that positive reinforcement is crucial to building green coalitions (Kelsey, 2014; Stokes, 2015; Laird and Stefes, 2009). The concept of positive reinforcement draws from the literature on path dependence where each step along a particular path produces consequences that increase the relative attractiveness of that path. The further along the path, the harder it is to reverse or change paths (Pierson, 2004). While path dependency is typically used to explain inertia, the insights can be used to consider how these types of processes might be used to reinforce a transition to a sustainable energy path. In doing so, it forces policymakers not only to consider the cost-benefit rationale for policies, which tend to dominate public policy choices, but also to consider which policy is more likely to entrench specific constituencies by increasing returns and positive feedbacks, and over time to expand that constituency (Levin *et al.*, 2012: 141).

Entrench and build existing interests

In this context, the first strategy for policymakers is to design policies that entrench and build existing interests. Targeted sector-specific policies, such as subsidies and tax rebates, that provide concentrated benefits to firms and households, as opposed to general economy-wide policies, are more easily entrenched over time. This is because the constituencies that benefit from them tend to become politically bound to the policy, making them active supporters and defenders of it (Holland *et al.*, 2014). Over time, such constituencies may come to support stricter climate and energy regulations, making far-reaching attempts to regulate energy more likely to succeed (Meckling *et al.*, 2015; Levin *et al.*, 2012).

Renewable tax credits for solar power described in Chapter 5 provide a good illustration. Since 2006, the solar industry in the US has benefitted from the ITC, which reduces federal income taxes for capital investments in solar systems on residential and commercial properties. In doing so, it has helped to entrench and build support for solar power. There are several reasons why. First, the ITC, in tandem with other policies, such as renewable portfolio standards, have facilitated a surge in solar power in the US, entrenching the strength of the industry, which is now valued at around $15 billion compared to $800 million a decade ago (Resch, 2014). Second, such policies have been successful because they have reinforced an existing market trend, namely the plunging cost of solar PV, which has fallen by over 80 per cent since 2008 (MIT, 2015: 79). Third, in doing so, solar tax credits have nurtured a political constituency in support of solar power by providing a concentrated financial benefit to solar corporations. As industry revenues have increased so has the industry's power to defend the ITC and advocate for other clean energy regulations. This is evident in the growing lobbying presence of industry associations, such as SEIA, and the establishment of new political coalitions, such as TUSK. In sum, targeted sector-specific policies, such as renewable tax credits, are likely to prove an effective strategy for policymakers seeking to build political support for further policy action.

Exploit industry divisions

As well as entrenching and building existing interests, policymakers should also seek to exploit divisions across the energy sector. This is because when incumbent industries are divided and or politically weak, green coalitions and networks will be easier to build. First, in following their commercial interests energy corporations invariably become entangled in inter-industry conflicts. These can be exploited by smart policies. For example, if the aim is to regulate oil, policies are more likely to succeed if they exploit the natural divisions between oil producers and oil refiners rather than taking on both industries together. As discussed in the contest over oil exports, independent oil refiners mobilised to oppose oil producers because of the uneven distributive impact of allowing exports. While the refiners were unsuccessful, they likely weakened the political strength of producers. Likewise, if the aim is to regulate coal, the natural divisions between coal producers that mine the coal and the electric utilities that burn it to generate electricity can be leveraged in the same way. Again, these inter-industry divisions were evident in both the contest over the Waxman–Markey bill and the Clean Power Plan. As I will discuss below, shifting the interests of the industry itself can exacerbate these natural divisions.

Second, and related, when exploiting divisions policymakers should target industries that are less able to mount a resistance campaign. Given that the capacity of business actors to shape policy outcomes will largely be a function of the resources business actors have at their disposal and their political legitimacy, it makes sense to target industries that have limited or declining resources and legitimacy. In the US the obvious example is the coal industry (Downie, 2017a). As discussed in Chapter 4, the US coal industry is in structural decline, with production and revenue falling steadily. In fact, by 2015, more than 50 firms representing 50 per cent of US coal production had filed for bankruptcy protection, including three of the largest coal corporations – Peabody Energy, Alpha Natural Resources and Arch Coal – whose share prices have plummeted (IEA, 2016b: 204; Kary *et al.*, 2016). Coal's decline has manifested in a shrinking lobbying presence, with key coalitions, such as ACCCE, reducing their operations, and with it the industry's capacity to resist regulations. In addition, the political legitimacy of coal is also in decline with a raft of non-state actors now targeting its social licence to operate (Ayling, 2017; Gunningham *et al.*, 2004), and polling in the US consistently showing Americans want less reliance on fossil fuels, especially coal (Ansolabehere and Konisky, 2014). Consequently, when exploiting industry divisions policymakers should target those sections of industry that are least able to politically resist first, rather than trying to take on the energy sector all at once.

Third, policymakers should also seek to exploit intra-industry divisions. This may be more difficult than the above approaches because such divisions are not always easy to detect, especially when business actors hedge their position. Nevertheless, one way to exploit such divisions is to target outliers. That is, corporations that form a preference opposed to the rest of the industry. Indeed outliers emerged in almost all policy contests, even when the industry took a near unanimous position. For example, coal producers uniformly rounded on the Obama administration over the Waxman–Markey bill and later the Clean Power Plan. Yet in both cases Rio Tinto was the only major coal producer to support these initiatives. Policymakers who target these corporations, especially when they are powerful actors, such as Rio Tinto, may be able to bring them on board to support a policy that the rest of the industry opposes, thereby creating political cover. Of course, in bringing business actors on board policymakers will need to remain wary of firms that hedge their position. In other words, though they may publicly support a policy they could also be indirectly opposing it. At the same time, encouraging outliers to withdraw from key industry coalitions will reduce the financial resources and political legitimacy of the coalition. If policymakers are fortunate, the withdrawal of one significant corporation may create a cascade of departures. Given that the behaviour of outliers tends to reflect specific institutional factors, policymakers would be well advised to monitor business actors that for instance, have a different home country or a unique history to the rest of the industry; two institutional factors that have been shown to produce outliers.

Shift existing interests

Policymakers should not only seek to entrench existing interests and exploit divisions, but also to shift interests and create wider coalitions and networks. Policy signals need to induce significant changes in industry investment and structure. If policymakers consistently target low-hanging fruit, they may find that once the fruit has been picked resistance from incumbent industries will not have shifted and future more stringent climate regulations will be equally hard to implement (Kelsey, 2014: 6). Policies need to send strong direct signals that will drive change, such as renewable portfolio standards that mandate utilities to supply a specific percentage of renewable energy, which can lead to structural configuration in the utility industry as utilities invest in renewable generation at the expense of coal-fired generation (Meckling *et al.*, 2015). In other words, it will shift their commercial interests toward policies that support renewable energy over fossil fuels.

Policymakers also need to send repeat signals that demonstrate to business actors that the regulatory landscape is changing. Over time this can lead to a tipping point in an industry, when a critical mass of affected corporations shifts away from opposing a specific regulatory initiative. When proposals for new regulations first emerge on the agenda business actors are likely to oppose, but repeated signals from policymakers create a perception within the industry that regulation is inevitable leading them to support the regulations, or hedge their position and attempt to shape the forthcoming regulation (Vormedal, 2011). In both cases, repeat signals have the effect of limiting resistance. This dynamic was evident in the coal and utility industries. Over the last two decades successive US administrations have instigated a series of initiatives to regulate coal. President Obama's attempt to introduce emissions trading via the Waxman–Markey bill was simply the latest attempt. As discussed in Chapter 4, for many utilities the motivation for joining USCAP, one of the key coalitions supporting the bill, was to shape the legislation because, according to one of the founders of USCAP, there was not a single USCAP CEO who did not consider future climate change regulation inevitable (Bartosiewicz and Miley, 2013: 26). The same dynamic was in play four years later with the Clean Power Plan. As executives and lobbyists in the coal and utility industries repeatedly pointed out, the successive attempts at regulation, whether successful or not, 'have created a market psychology that coal is never coming back, and that is the most important impact … the feeling is that there is no future for coal' (Interview 49). The result was that many corporations began to limit their resistance as they looked to re-structure their business models for a new regulatory environment.

Resistance

Policymakers not only need to build coalitions that can overcome the resistance from incumbent fossil fuels industries, they also need to pursue policies that limit resistance. As other scholars have pointed out, incumbent industries in the US will mobilise against clean energy policies that threaten their short-term profitability and long-term existence. Indeed, there is a greater tendency to do so in adversarial political systems like the US, than in more cooperative systems like Europe (Hess, 2014: 279). Consequently, policymakers in the US will need to identify and design policies that limit resistance.

First, policies should seek to limit negative feedbacks. The most obvious way is to eliminate policies that build and entrench existing fossil fuel interests. Fossil fuel subsidies are a case in point. The OECD estimates that governments around the world spend an estimated $160–$200 billion per year subsidising fossil fuels (OECD, 2015).[2] Attempts to eliminate subsidies will be more likely to succeed when they take advantage of existing market conditions. For instance, the plunge in the global price of oil between 2014 and 2016 presented a unique opportunity for policymakers to abolish support for the consumption of fossil fuels. At the same time, policymakers should resist pressure to support oil and gas corporations in their exploration and production activities.

Policymakers should also avoid making trade-offs that could result in negative feedbacks. For example, in the contest over the Waxman–Markey bill, the Obama administration agreed to include US$1 billion in funding to support CCS. CCS had been actively promoted by both industries including many of the most powerful coal producers and coal-fired utilities, such as Peabody Energy, Southern Company, and American Electric Power (Interviews 6 and 46). While such large public subsidies for CCS technology, though they may have great potential to reduce emissions, and were arguably helpful in garnering support for the bill, can also work to lock in decades of further coal generation by providing a justification for its ongoing exploitation, which may help to explain Southern Company's continuing opposition to the Clean Power Plan. Policymakers will need to weigh carefully the trade-offs of such policies, especially in cases where the technologies have consistently failed to deliver, as is the case with CCS technology (Downie and Drahos, 2015).

Second, policymakers need to design policies that limit the avenues of resistance. This will be especially important when policymakers are attempting to build and entrench existing interests. In other words, when coalitions in support of clean energy policies are weak and incumbent fossil fuel industries resisting are comparatively strong. For example, coal producers conceded that it was easier to lobby against the Waxman–Markey bill than the Clean Power Plan, because there are more potential veto points on legislation that has to pass through Congress than regulations, which do not. In this context, non-legislative avenues may prove easier to pursue than legislative avenues to the extent that they limit potential veto points. Previous research has pointed to executive orders, the courts, and voter referenda as potential pathways to achieve environmental outcomes in the face of US Congressional gridlock (Klyza and Sousa, 2008).

Finally, policies can be framed in ways that make them harder to resist. There are two dimensions to this. First, policies that are tied to existing frames or principles that have normative appeal will be harder to oppose. Indeed, proponents of clean energy policies could take a leaf out of the oil and gas industry playbook, which was extremely successful in framing the solution to export restrictions on oil and gas as 'free trade'. As discussed in Chapter 3, because the established frame of free trade had widespread support in Congress, opponents in the petrochemical industry and refining industry found it difficult to resist. Second, frames that encourage business support across industries, and even across social movements, will also be more difficult to resist, especially given the tendency of incumbent industries to frame environmental policies as harmful to business and the economy (Hess, 2014: 279; Brown and Hess, 2016: 979). Continuing with the oil and gas example, the frame of free trade not only had normative appeal in Congress, but it also encouraged support from across the business community, including from powerful business associations, such as the COC and NAM, which as discussed, were unlikely to resist any policy that could be construed as anti-free trade.

In summary, implementing policies that promote clean energy, as history has shown, will not be easy. Indeed, they are unlikely to succeed in the absence of green coalitions that can overcome the resistance of incumbent fossil fuel industries. In this context, the insights of the empirical analysis of business actors in the US energy sector, suggest three specific strategies for policymakers. First, entrench and build existing interests in support of clean energy policy via targeted sector specific policies. Second, exploit inter-industry divisions with smart policies that ideally target politically weak industries that will be less able to mount a resistance campaign, and exploit intra-industry divisions by seeking to bring on board outliers that support policies that the rest of an industry opposes. Third, shift existing interests with policies that induce changes in industry investment and structure, such as renewable energy targets. Policies signals not only need to be direct, but they need to be repeated to create the impression that reform is inevitable. Further, such policies will work best when they are designed in such ways that limit the avenues for resistance.

It is also important to note that the order that these strategies have been considered in is not accidental, the sequencing of policies matter. Climate policies should first lock-in policies, however small and targeted, when there is a window of opportunity to do so. Sector-specific policies that provide concentrated benefits will over time entrench support for future rounds of clean energy policies including more difficult interventions that seek to shift interests and restructure industries, and ultimately economy-wide measures, such as emissions trading, which are more efficient at reducing greenhouse gas emissions (Meckling *et al.*, 2015).

Business strategies in the Trump era

This chapter has so far focussed on what policymakers can do to achieve a clean energy transition, yet in the wake of Donald Trump's election it is also worth considering the lessons for business actors in emerging renewable industries. In other words, what can 'green' business actors do in the absence of attempts by federal policymakers to advance an energy transition? And, given the current political environment, what can such business actors do in response to attempts to reverse course on existing clean energy policies? While the election of President Trump will generate greater state and local engagement on climate and energy policy, much like the election of President George W. Bush did before it (Selin and VanDeveer, 2009), the importance of the federal level necessitates that renewable industries do not simply vacate the field to incumbent industries.

Before turning to potential business strategies, it is helpful to briefly outline the changes in climate and energy policy under the Trump administration. As a presidential candidate, Trump campaigned to reverse course on many of President Obama's signature climate and energy initiatives, which since assuming office in January 2017 he has now begun to implement. At the international level, Trump announced in June in 2017 that he would withdraw the US from the Paris Agreement, which President Obama had strongly advocated for during his second term in office (The White House, 2017c). At the national level, President Trump has arguably gone further with a series of executive orders designed to roll back key initiatives, including the Clean Power Plan. Indeed, Scott Pruitt, the former Attorney General of Oklahoma, and leader of the Republican Attorneys General Association, who as discussed in Chapter 4 helped to lead resistance to the Clean Power Plan in the courts, was appointed as the first administrator of the EPA under Trump, tasked with dismantling the Clean Power Plan, though he has since been forced to resign over allegations of corruption (Davenport *et al.*, 2018). In addition, the White House has lifted the moratorium on coal leasing on federal lands, reversed regulations on methane emissions, and put an end to restriction on oil drilling in Arctic waters, among other initiatives (The White House, 2017a, 2017b).

In this context, it is worth considering the lessons of the preceding analysis not only for policymakers, but also for business actors in the emerging renewable industries. As discussed in Chapter 2, the effectiveness of business strategies and the capacity of business actors to shape policy outcomes to a large extent will be a function of the resources business actors have at their disposal and their political legitimacy. For example, this may include their financial resources, political, economic, and technical expertise, and it may include the extent to which business actors are considered appropriate within the societies in which they operate.

Against this backdrop several strategies stand out. First, business actors in renewable industries, such as wind and solar, need to continue to mobilise coalitions that build their resources and legitimacy. This is critical given the unequal capacity of incumbent fossil fuel industries relative to emerging renewable industries. Indeed, throughout the policy contests, renewable industries had more limited financial resources than their fossil fuel counterparts, for lobbying and other activities, and at the same time, they were often seen as outliers in the business community. In other words, they did not have the same political legitimacy, as evident in the policy positions taken by key business associations, which were often at odds with the preferences of firms in these industries.

Accordingly, one approach for renewable industries is to maintain and strengthen coalitions with firms in the technology sector, manufacturing sector and retail sector, which can provide much needed legitimacy in the eyes of policymakers in the Trump administration, and, of course, resources too. In the much the same way as solar firms sought to mobilise firms, such as Google and Starbucks, in the Obama era, as discussed above, it is arguably now more imperative in the Trump era to do the same if green business actors are to have any chance of sustaining existing federal policies that encourage clean energy. A related approach is to reduce the resources and legitimacy of existing business coalitions that oppose policies in support of renewable energy. For example, encouraging prominent US firms to withdraw from a business coalition, such as the NAM, should it support the repeal of subsidies for solar power, could potentially create a cascade of departures thereby weakening support in the business community for new regulatory initiatives under the Trump administration.

Second, green business actors need to defend those policies that entrench and build existing interests in renewable energy industries. In other words, policies that nurture a political constituency in support of renewable energy by providing concentrated financial benefits to firms and households. As discussed, tax credits for wind and solar have played this role facilitating growth in both industries, which in turn has enhanced the capacity of these industries to defend the PTC and the ITC as evident in their growing lobbying presence. Of course, renewable tax credits are only one of a number of policies that have played this role. Net metering regulations are another example. Under net metering, utilities compensate households and commercial premises with solar power for the electricity they generate and feed back into the grid at the retail price (MIT, 2015: 219). Because the benefits of net metering increase over time and accrue not just to solar firms, but households too, it entrenches as well as expands support for solar power (Interview 41). Recent studies on net-metering regulations in the US confirm this, as does evidence from other nations, such as Germany, which have shown how feed-in tariffs that provide similar concentrated benefits, strengthened the political clout of renewable energy industries (Jacobsson and Lauber, 2006, Stokes, 2015). While largely regulated at the state level in the US, there has been the ongoing threat of federal intervention to curb these policies, which have proven effective at building political support for further policy action.

The question then is how to defend such policies, especially given that incumbent fossil fuel industries may see an opportunity to scuttle these policies with President Trump in the White House. One way is to mobilise the political coalitions described above, because as other empirical studies have concluded, sustained political support for renewable industries is likely to be vital to achieving an energy transition (Stokes and Breetz, 2018). However, should traditional lobbying campaigns fail in this environment to sustain these policies, green business actors may have to hedge their position. As discussed in Chapter 6, in the wind and solar industries firms fearing that they would be unable to stop the expiration of the PTC and the ITC, advocated for these tax credits to be phased-out rather than expire immediately. In other words, they hedged their position by shaping the rules in ways that bought them more time for the industry to grow and nurture supporting political constituencies.

Research implications

With attempts to regulate the energy sector frequently resisted by business actors, it makes sense for policymakers to consider not just the economic rationale for a policy, but also whether it builds support for more efficient policies. In this context, more work needs to be done to understand how and why incumbent fossil fuel industries influence policy outcomes. At the descriptive level, the policy positions of key industries in energy the sector need to be distinguished from the public rhetoric for upcoming policy contests under the Trump administration. This will help policymakers to identify cleavages within and between industries, including potential outliers. A systematic mapping of the types of business strategies will also be useful. For example, the empirical evidence presented here suggests that the forms of collective action business take vary considerably from formal coalitions, such as industry associations, to informal ad hoc coalitions. Different forms of collective organisation are likely to have different impacts on policy outcomes.

Analytically, identifying the policy positions of key industries requires a better understanding of hedging behaviour. The empirical evidence presented here indicates that in the US energy sector industries hedge their position and shape regulations when they believe they cannot stop the proposed regulations, and when they believe that they cannot sustain regulations with their support. Future research should consider the precise conditions under which hedging positions are adopted. A better understanding of this behaviour will be crucial to scholars seeking to explain specific policy outcomes and policymakers seeking to build momentum for regulations so that firms believe they are inevitable and cannot be stopped. This would also be beneficial to policymakers seeking to enlist corporate allies in the energy sector, because in many cases hedging may obscure the true position of firms.

Similarly, distinguishing between different types of business strategies, requires further analytical work to establish the precise conditions under which business select specific strategies. This book suggests a number of areas where this could be developed. This could include examining the conditions under which traditional industry associations become the command centre of business campaigns tying together networks of actors. In the same vein future studies should identify the factors that explain why business actors establish ad hoc coalitions, because the evidence presented here indicates that such coalitions are particularly appealing to business actors looking to resist attempts at regulation. In addition, further analytical work is needed to conceptualise the role that coalitions more generally can play in building the legitimacy of emerging industries, such as the wind and solar industries.

Finally, and importantly, the findings presented here are likely to be generalisable to cases that share similar characteristics, such as policy contests that take place in the US, at the Federal level, and involve large firms in established industries. For example, it can be expected that this analytical framework would be useful for understanding how and why business actors in the pharmaceutical or automotive industries shape policy outcomes at the Federal level in the US. However, further empirical work is needed to consider whether the findings of this book are the same in different contexts. For example, in the US context, would the same insights be applicable to industries populated by small firms contesting policy at the state level?

Similarly, while the findings are likely to be readily applicable in other jurisdictions with similar political systems, such as Australia, Canada, and the United Kingdom, there will be differences. For example, Europe has very different governmental structures and legal traditions than the US, which may mean that in some cases adversarial strategies are less appropriate (Kagan, 2007). Further, the strategies identified in this chapter are likely to be less applicable in nations with dramatically different political systems, such as China and Russia. Understanding these differences is important, especially given that these nations represent some of the largest producers and consumers of energy in the world.

Climate implications

As this book has shown, regulating the energy sector will not be easy. Policy outcomes in every industry from oil and gas, to coal, to utilities, to wind and solar, are fiercely contested and business actors have developed complex means by which they seek to exercise influence over the policy process as they pursue their commercial interests. While the focus of this chapter has been on what policymakers should do in such circumstances, it is worth pausing to consider the implications of what is happening now. Specifically, the effect of the transformations currently taking place in the

US oil and gas sector, coal sector, and renewables sector, and what they mean for achieving a clean energy transition. In doing so, I will also touch briefly on the potential impact of President's Trump's administration in these sectors, to the extent that it can so far be discerned.

First, the shale revolution and the subsequent rapid increase in oil and gas production is incompatible over the long run with the aim of keeping global temperatures 'well below' 2°C. As discussed, this requires that a third of the world's oil reserves and half the world's gas reserves be left in the ground. More specifically, it requires that the US does not burn 9 per cent of its oil reserves and a further 6 per cent its gas reserves (McGlade and Ekins, 2015: 189). Yet the US like the rest of the world remains off track when it comes to the oil and gas industry. Many governments and corporations continue to invest in long-lived oil and gas infrastructure despite their dangerous contribution to climate change. Policy decisions that promote these commodities for commercial gain are only likely to compound the problem. For example, lifting restrictions on oil and gas exports will encourage further exploration, even when the evidence shows that the climate is in crisis from burning existing fossil fuel reserves. This is likely to be exacerbated by the Trump administration, which is considering lifting existing restrictions on oil and gas exploration, including in the Arctic National Wildlife Refuge (Friedman, 2017).

Second, and in contrast, the structural decline of the coal industry, which was accelerated by the Obama administration's attempts to limit emissions from the coal, will have positive implications for global attempts to reduce greenhouse gas emissions. As noted, if the worst impacts of climate change are to be avoided, almost 90 per cent of global coal reserves must be left in the ground. In the US, which has the world's largest recoverable coal reserves, this translates to 95 per cent of domestic reserves (McGlade and Ekins, 2015). Consequently, any measures that the US takes to limit coal production will have positive implications. While the measures taken by the Obama administration were not enough, the critical point is that they discouraged coal production, in contrast to what has occurred in the oil and gas sector. A report by Goldman Sachs stated that 'investment in new coal-fired capacity can be discouraged simply by the risk of new regulations, such as the rules under consideration by the EPA' (Lelong *et al.*, 2013: 20).

However, the election of President Trump has no doubt shifted some sentiment within the coal industry and the new administration has begun rolling back a series of existing rules designed to reduce coal consumption. For example, in addition to stopping the implementation of the Clean Power Plan, the Trump administration has overturned a freeze on new coal leases on public lands, methane reporting requirements and anti-dumping rules for coal, with more proposals in the pipeline (Popovich and Albeck-Ripka, 2017). Nevertheless, even with these initiatives it remains unlikely to be able to reverse coal's structural decline, especially in the face of historically low gas prices (Galik *et al.*, 2017).

Third, in the US renewable energy, and especially wind and solar power, are thriving. As discussed in Chapter 5, in 2015 renewables represented 10 per cent total US energy consumption, its highest share since the 1930s, when wood was a primary fuel source (EIA, 2015c). According to the IEA, if the world is to have a 50 per cent probability of limiting warming to 2°C, the share of renewables in global total primary energy demand must rise to almost 30 per cent by 2040, up from 14 per cent today. For the US, this means the share of renewables in total primary energy demand must quadruple (IEA, 2015b: 347–348). Wind and solar will have to play a big part and government support schemes, such as renewable tax credits, will remain important. While it is too early to discern whether the Trump administration will seek to roll back the extensions to the PTC and the ITC agreed to in 2015 as part of broader tax reform, should they do so, this would likely slow, though not stall growth in these industries.

Further, the IEA estimates that in the next decade the US will require $255 billion per year in energy supply investment, in large part due to the need to replace ageing power plants and integrate renewables into the electricity grid (IEA, 2015a: 46). Yet recent studies have shown that for a 2°C target to be met none of this investment can be made into fossil electricity infrastructure, without the early stranding of such assets or carbon capture, given that investments made today are likely to be operating and contributing greenhouse gas emissions for decades to come. For policymakers therefore, the focus must be as much on future investments as it is on the operation of existing assets (Pfeiffer *et al.*, 2016).

In summary, the US is an energy superpower and the transformations taking place in its energy sector will be critical to determining whether the world can achieve a clean energy transition. Yet it is hard to imagine such a transition taking place without overcoming the political resistance of incumbent fossil fuel industries. Indeed, resistance from industries, such as oil, gas and coal, in particular, could derail efforts to regulate energy in a sustainable fashion. By examining business behaviour in contemporary policy contests across the US energy sector, this book aims to improve our understanding of how and why business actors behave in a sector that has long been overlooked by scholars. Building on the existing literature and the empirical data analysed in this book, I hope to have provided not only theoretical insights for scholars, but also practical insights for policymakers who are seeking to ensure that a clean energy transition is not the hope of future decades, but the reality of this one.

Notes

1 Sections of this chapter are reprinted from Christian Downie 2017a, Business actors, political resistance, and strategies for policymakers. *Energy Policy*, 108, 583–592 © (2017), with permission from Elsevier.
2 The figures are for OECD countries, plus six partner countries, namely, Brazil, China, India, Indonesia, Russia, and South Africa.

References

Adcox, S. 2017. Billions lost in nuclear power projects, with more bills due. *AP News*, 5 August.

Adelman, D. E. A. S. and David B. 2017. Ideology vs. interest group politics in US energy policy. *North Carolina Law Review*, 95, 339–414.

Adler, J. 2016. Supreme Court puts the brakes on the EPA's Clean Power Plan. *Washington Post*.

AEA. 2015a. *End wind welfare* [Online]. American Energy Alliance. Available: http://americanenergyalliance.org/take-action/endwindwelfare/ [accessed 7 September 2015].

AEA. 2015b. *PTC Elimination Act protects American families* [Online]. Washington DC: American Energy Alliance. Available: http://americanenergy alliance.org/2015/04/22/ptc-elimination-act-protects-american-families/ [accessed 17 October 2015].

Aggarwal, V. K. 2001. Corporate market and nonmarket strategies in Asia: A conceptual framework. *Business and Politics*, 3, 89–108.

Aklin, M. and Urpelainen, J. 2013. Political competition, path dependence, and the strategy of sustainable energy transitions. *American Journal of Political Science*, 57, 643–658.

ALEC. 2015. *Act requiring approval of state plan implement EPA's carbon guidelines.* [Online]. American Legislative Exchange Council. Available: www.alec.org/model-policy/act-requiring-approval-state-plan-implement-epas-carbon-guidelines/ [accessed 12 February 2016].

Allison, G. 1971. *Essence of decision: Explaining the Cuban Missile Crisis.* New York: Harper Collins.

American Electric Power. 2009. AEP's Position on Climate Legislation Is Clear: A Message from AEP Chairman, President & CEO Michael G. Morris. Press Release, 14 September 2009. Available: https://grist.files.wordpress.com/2010/09/mikemorriswaxman-markeystatement.pdf [accessed 9 June 2015].

Anon. 2013. Wyden blasts exports again. *LNG Intelligence.* 11 January.

Anon. 2015. Wind power industry seeks phaseout of tax credit amid Republican scrutiny. *FARS News Agency*, 17 January.

Ansolabehere, S., De Figueiredo, J. M., and Snyder, J. M. 2003. Why is there so little money in US politics? *The Journal of Economic Perspectives*, 17, 105–130.

Ansolabehere, S. and Konisky, D. M. 2014. *Cheap and clean: How Americans think about energy in the age of global warming.* Cambridge, MA: MIT Press.

API. 2014a. *Erik Milito's remarks at press briefing on LNG exports to US allies* [Online]. American Petroleum Institute. Available: www.api.org/news-and-media/testimony-speeches/2014/erik-milito-press-briefing-on-lng-exports-to-us-allies [accessed 25 August 2016].

API. 2014b. Understanding natural gas markets. American Petroleum Institute. Available: www.api.org/~/media/Files/Oil-and-Natural-Gas/Natural-Gas-primer/Understanding-Natural-Gas-Markets-Primer-High.pdf [accessed 14 February 2015].

API. 2015. Every major study agrees: Crude oil exports would put downward pressure on US gasoline prices. American Petroleum Institute. Available: www.api.org/~/media/files/policy/exports/economic-studies-crude-oil-exports.pdf [accessed 27 October 2015].

AWEA. 2013. *American Wind Energy Association names Tom Kiernan CEO* [Online]. Washington DC: American Wind Energy Association. Available: www.awea.org/resources/press-releases/2013/american-wind-energy-association-names-tom-kiernan [accessed 4 August 2013].

AWEA. 2014. *US Wind Industry annual market report*. Washington DC: American Wind Energy Association.

AWEA. 2015a. *US Wind Industry annual market report*. Washington DC: American Wind Energy Association.

AWEA. 2015b. *US Wind Industry leaders praise multi-year extension of tax credits*. Washington DC: American Wind Energy Association.

AWEA. n.d. *Polling on wind energy policy* [Online]. Washington DC: American Wind Energy Association. Available: www.awea.org/MediaCenter/content.aspx?ItemNumber=9071 [accessed 12 August 2016].

Ayling, J. 2017. A contest for legitimacy: The Divestment movement and the fossil fuel industry. *Law & Policy*, 39: 349–371. doi: 10.1111/lapo.12087

Ayling, J. and Gunningham, N. 2015. Non-state governance and climate policy: The fossil fuel divestment movement. *Climate Policy*, 17(2), 131–149, doi: 10.1080/14693062.2015.1094729

Barley, S. R. 2010. Building an institutional field to corral a government: A case to set an agenda for organization studies. *Organization Studies*, 31, 777–805.

Barnet, R. and Muller, R. 1974. *Global reach: The power of the multinational corporations*. New York: Simon and Schuster.

Barnett, M. and Duvall, R. (eds) 2005. *Power in global governance*. Cambridge: Cambridge University Press.

Barradale, M. J. 2010. Impact of public policy uncertainty on renewable energy investment: Wind power and the production tax credit. *Energy Policy*, 38, 7698–7709.

Bartosiewicz, P. and Miley, M. 2013. *The too polite revolution: Why the recent campaign to pass comprehensive climate legislation in the United States failed*. New York: Columbia University.

Baumgartner, F. R., Berry, J. M., Hojnacki, M., Kimball, D., and Leech, B. 2009. *Lobbying and policy change: Who wins, who loses, and why*. Chicago, IL: University of Chicago Press.

Baumgartner, F. R. and Jones, B. D. 1991. Agenda dynamics and policy subsystems. *The Journal of Politics*, 53, 1044–1074.

Beach, D. and Pedersen, R. 2013. *Process-tracing methods: Foundations and guidelines*. Ann Arbor, MI: University of Michigan Press.

Beach, W. W., Lieberman, B., Campbell, K., and Kreutzer, D. W. 2009. *Son of Waxman–Markey: More politics makes for a more costly bill* [Online]. The Heritage Foundation. Available: www.heritage.org/research/reports/2009/05/son-of-waxman-markey-more-politics-makes-for-a-more-costly-bill [accessed 25 August 2016].

Bell, J. 2013. AEP cautiously optimistic about President Obama's climate plans. *Columbus Business First*, 27 June 2013.

Bell, J. 2014. The time has come to lift the oil export ban. 10 January. Available: www.ipaa.org/2014/01/10/the-time-has-come-to-lift-the-oil-export-ban/ [accessed 14 February 2015].

Bennett, A. 2007. Case study methods: Design, use, and comparative advantages. *In:* Sprinz, D. and Wolinsky-Nahmias, Y. (eds) *Models, numbers & cases: Methods for studying international relations.* Ann Arbor, MI: The University of Michigan Press.

Bernstein, S. 2011. Legitimacy in intergovernmental and non-state global governance. *Review of International Political Economy*, 18, 17–51.

Berry, J. M. 2002. Validity and reliability issues in elite interviewing. *PS: Political Science and Politics*, 35, 679–682.

Betsill, M. 2008. Reflections on the analytical framework and NGO diplomacy. *In:* Betsill, M. and Corell, E. (eds) *NGO diplomacy: The influence of non-governmental organizations in international environmental negotiations.* Cambridge: MIT Press.

Biersteker, T. J. 1980. The illusion of state power: Transnational corporations and the neutralization of host-country legislation. *Journal of Peace Research*, 17, 207–221.

Blau, G. 2015. *IBISWorld industry report: Petrochemical manufacturing in the US.* Melbourne: IBISWorld.

Blyth, M. and Matthijs, M. 2017. Black swans, lame ducks, and the mystery of IPE's missing macroeconomy. *Review of International Political Economy*, 24, 203–231.

BNEF. 2015. *Medium-term outlook for US power.* New York: Bloomberg New Energy Finance.

Boddewyn, J. J. and Brewer, T. L. 1994. International-business political behavior: New theoretical directions. *The Academy of Management Review*, 19, 119–143.

Boden, T. A., Marland, G., and Andres, R. J. 2017. *Global, regional, and national fossil-fuel CO_2 emissions.* Oak Ridge National Laboratory. Oak Ridge, TN: Carbon Dioxide Information Analysis Center, US Department of Energy.

Bowman, S. 1996. *The modern corporation and American political thought: Law, power and ideology.* University Park, PA: The Pennsylvania State University Press.

Braithwaite, J. and Drahos, P. 2000. *Global business regulation.* Cambridge: Cambridge University Press.

Bravender, R. 2014. *House Republican compares climate, coal rules to 'terrorism'* [Online]. 28 July. Available: www.eenews.net/greenwire/2014/07/28/stories/1060003625 [accessed 4 August 2014].

Broder, J. M. 2009a. EPA moves to curtail greenhouse gas emissions. *New York Times*, 30 September 2009.

Broder, J. M. 2009b. House passes bill to address threat of climate change. *New York Times*, 26 June 2009.

Brown, K. P. and Hess, D. J. 2016. Pathways to policy: Partisanship and bipartisanship in renewable energy legislation. *Environmental Politics*, 25, 971–990.

Brown, S. P. A., Mason, C., Krupnick, A., and Mares, J. 2014. Crude behaviour: How lifting the export ban reduces gasolin prices in the United States. Resources for the Future. Available: www.rff.org/files/sharepoint/WorkImages/Download/RFF-IB-14-03-REV.pdf [accessed 15 June 2014].

Burris, S., Drahos, P., and Shearing, C. 2005. Nodal governance. *Australian Journal of Legal Philosophy*, 30, 30–58.

C2ES. 2009. In brief: What the Waxman–Markey bill does for coal. Center for Climate and Energy Solutions. Available: www.c2es.org/document/in-brief-what-the-waxman-markey-bill-does-for-coal/ [accessed 4 August 2015].

Capitol Tax Partners. 15 April 2015. *RE: Letter to Senate Committee on Finance.* Type to Thune, J., Cardin, B. L., Heller, D., and Bennet, M. Available: www.finance.senate.gov/imo/media/doc/Capitol%20Tax%20Partners1.pdf

Cardwell, D. 2015. Worry for solar projects after end of tax credits. *New York Times*, 25 January 2015.

Carroll, J. and Tobben, S. 2016. First US oil export leaves port; marks end to 40-year ban. *Bloomberg*, 1 January 2016.

Carvill, P. 2015. Obama seeks to extend ITC permanently. *PV Magazine*, 3 February 2015.

Cass, L. 2007. The indispensable awkward partner: The United Kingdom in European climate policy. *In:* Harris, P. (ed.) *Europe and global climate change: Politics, foreign policy and regional cooperation.* Cheltenham: Edward Elgar.

Ceres Policy Network. n.d. *Business for Innovative Climate and Energy Policy (BICEP)* [Online]. Ceres. Available: www.ceres.org/networks/ceres-policy-network [accessed 3 June 2016].

Chapman, K. and Dodge, C. 2009. Obama plan has $79 billion from cap-and-trade in 2012 (update3). *Bloomberg*, 26 February 2009.

Chediak, M. 2017. *Southern's clean coal experiment ends with Mississippi order* [Online]. 7 July: Bloomberg. Available: www.bloomberg.com/news/articles/2017-07-06/southern-s-clean-coal-experiment-officially-dead-on-state-order [accessed 20 September 2017].

Chediak, M. and Weber, H. 2014. Polar vortex emboldens industry to push old coal plants. *Bloomberg*, 11 March 2014.

Chemnick, J. 2015. Republican statehouse gains load the dice for anti-EPA bills. *E&E News*, 12 January 2015.

Clapp, J. and Fuchs, D. 2009. Agrifood corporations, global governance, and sustainability: A framework for analysis. *In:* Clapp, J. and Fuchs, D. (eds) *Corporate power in global agrifood governance.* Cambridge: MIT Press.

Clapp, J. and Meckling, J. 2013. Business as a global actor. *In:* Falkner, R. (ed.) *The Handbook of global climate and environmental policy.* West Sussex: John Wiley & Sons.

Cocklin, J. 2015. ANGA taking natural gas know-how to API in merger deal. *Natural Gas Intel*, 18 November 2015.

Collingwood, V. 2006. Non-governmental organisations, power and legitimacy in international society. *Review of International Studies*, 32, 439–454.

Commodities Now. 2013. Increasing oil and gas production not at odds with lowering usage. *Commodities Now*, 13 December 2013.

Congress.gov. 2014. *H.R.4349 – Crude Oil Export Act* [Online]. Available: www.congress.gov/bill/113th-congress/house-bill/4349 [accessed 3 June 2016].

Copley, M. 2013. Eyeing legislative vehicles, lawmakers for and against wind PTC lobby tax reformers. *SNL Energy Electric Utility Report*, 25 November.

Copley, M. 2014. Ahead of tax-extenders bill, wind advocates ramp up lobbying effort. *SNL Renewable Energy Weekly*, 4 April 2014.

Copley, M. 2015a. At a crossroads, wind lobbyists plan to attack unfriendly lawmakers. *SNL Power Week Canada*, 25 May 2015.

Copley, M. 2015b. Wind industry betting on PTC extension. *SNL Energy Finance Daily*, 16 October 2015.

Corell, E. and Betsill, M. 2008. Analytical framework: Assessing the influence of NGO diplomats. *In:* Corell, E. and Betsill, M. (eds) *NGO diplomacy: The influence of nongovernmental organizations in international environmental negotiations.* Cambridge: The MIT Press.

Crooks, E. 2013. Opposition mounts to US gas exports. *Financial Times*, 25 March 2013.

Crowley, K. 2017. Up and down with climate politics 2013–2016: The repeal of carbon pricing in Australia. *WIREs Clim Change*, 8: e458. doi: 10.1002/wcc.458

CRP. 2009. *US climate action partnership* [Online]. Center for Responsive Politics. Available: www.opensecrets.org/lobby/clientsum.php?id=D000058047&year=2009 [accessed 12 January 2016].

CRP. 2014. *Coal mining: Lobbying, 2014* [Online]. Center for Responsive Politics. Available:www.opensecrets.org/industries/lobbying.php?cycle=2014&ind=e1210 [accessed 25 August 2016].

CRP. 2015a. *American Wind Energy Assn* [Online]. Washington DC: The Center for Responsive Politics. Available: www.opensecrets.org/ lobby/clientsum. php?id=D000024034&year=2015 [accessed 17 July 2016].

CRP. 2015b. *Dow chemical* [Online]. Center for Responsive Politics. Available: www.opensecrets.org/lobby/clientsum.php?id=D000069022&year=2011 [accessed 4 August 2014].

CRP. 2015c. *Oil & gas: Industry profile: Summary, 2015* [Online]. Center for Responsive Politics. Available: www.opensecrets.org/lobby/indusclient.php? id=E01&year=2015 [accessed 25 August 2016].

CRP. 2016a. *America's natural gas alliance: Lobbying totals, 1998–2016* [Online]. Center for Responsive Politics. Available: www.opensecrets.org/orgs/lobby. php?id=D000046794 [accessed 25 August 2016].

CRP. 2016b. *American coalition for clean coal electricty* [Online]. Center for Responsive Politics. Available: www.opensecrets.org/lobby/clientsum.php?id=D000 046880&year=2012 [accessed 18 July 2016].

CRP. 2016c. *Coal mining: Industry profile: Summary, 2016* [Online]. Center for Responsive Politics. Available: www.opensecrets.org/lobby/induscode.php?id= E1210 [accessed 25 August 2016].

CRP. 2016d. *Edison Electric Institute: Client profile: Summary 2016* [Online]. Center for Responsive Politics. Available: www.opensecrets.org/lobby/clientsum. php?id=D000000297 [accessed 25 August 2016].

CRP. 2016e. *Electric utilitites: Industry profile: Summary, 2016* [Online]. Center for Responsive Politics. Available: www.opensecrets.org/lobby/indusclient. php?id=E08 [accessed 25 August 2016].

CRP. 2016f. *First solar: Spending by cycle* [Online]. Center for Responsive Politics. Available: www.opensecrets.org/pacs/lookup2.php?strID=C00489534&cycle =2014 [accessed 4 August 2017].

CRP. 2016g. *Lobbying ranked by sectors* [Online]. Center for Responsive Politics. Available: www.opensecrets.org/lobby/top.php?showYear=2016&indexType=c [accessed 9 May 2017].

CRP. 2016h. *Solar Energy Industries Assn: Spending by cycle* [Online]. Center for Responsive Politics. Available: www.opensecrets.org/pacs/lookup2.php?strID=C00421982&cycle=2016 [accessed 12 April 2017].

CRP. 2016i. *Solar Energy Industries Association client profile: Summary, 2016* [Online]. Center for Responsive Politics. Available: www.opensecrets.org/lobby/clientsum.php?id=D000029109 [accessed 25 August 2016].

CRP. 2016j. *SolarCity corp: Spending by cycle* [Online]. Center for Responsive Politics. Available: www.opensecrets.org/pacs/lookup2.php?cycle=2014&strID=C00520569 [accessed 2 May 2017].

CRP. 2017. *NextEra energy, client profile summary* [Online]. Center for Responsive Politics. Available: www.opensecrets.org/lobby/clientsum.php?id=D000000321 [accessed 4 August 2017].

CRP. n.d. *Lobbying totals, 1998–2016* [Online]. Washington DC: Center for Responsive Politics. Available: www.opensecrets.org/orgs/ lobby.php?id=D000000368 [accessed 8 February 2017].

Culpepper, P. D. 2010. *Quiet politics and business power: Corporate control in Europe and Japan.* New York: Cambridge University Press.

Culpepper, P. D. 2015. Structural power and political science in the post-crisis era. *Business and Politics*, 17, 391–409.

Dahl, R. 1961. *Who governs? Democracy and power in an American city*, New Haven, CT: Yale University Press.

Davenport, C. 2014a. Justices back rule limiting coal pollution. *New York Times*, 29 April 2014.

Davenport, C. 2014b. Republicans to investigate environmental group's influence on carbon rule. *New York Times*, 10 October 2014.

Davenport, C. 2015. Later deadline expected on Obama's climate plan. *New York Times*, 28 July 2015.

Davenport, C., Friedman, L., and Haberman, M. 2018. EPA chief Scott Pruitt resigns under a cloud of ethics scandals. *New York Times*, 5 July 2018.

Davenport, C. and Hirschfeld Davis, J. 2015. Move to fight Obama's climate plan started early. *New York Times*, 3 August 2015.

Delaney, K. 2007. Methodological Dilemmas and opportunities in interviewing organizational elites. *Sociology Compass*, 1, 208–221.

Dempsey, L. 2013. Banning LNG exports will hurt jobs and economy. *Shopfloor.org* [Online]. 15 January. Available: www.shopfloor.org/2013/01/banning-lng-exports-will-hurt-jobs-and-economy/27328/ [accessed 12 April 2014].

Desombre, E. R. 1995. Baptists and bootleggers for the environment: The origins of United States unilateral sanctions. *The Journal of Environment & Development*, 4, 53–75.

DiMaggio, P. J. and Powell, W. W. 1983. The iron cage revisited: Institutional isomorphism and collective rationality in organizational fields. *American Sociological Review*, 48, 147–160.

Dlouhy, J. A. 2015a. API launches first oil export ads as Congress nears vote. 9 September 2015. *Fuel fix* [Online]. Available: http://fuelfix.com/blog/2015/09/09/api-launches-first-oil-export-ads-as-congress-nears-vote/ [accessed 12 January 2016].

Dlouhy, J. A. 2015b. Senate panel advances oil exports in near party-line vote, signalling trouble ahead for bill. *Houston Chronicle*, 1 October 2015.

DoE. 2015. Chapter 1: Energy challenges. *Quadrennial Technology Review: An Assessment of Energy Technologies and Research Opportunities*. Washington DC: Department of Energy.

DoE. 2016. *Wind technologies market report*. Department of Energy: https://energy. gov/sites/prod/files/2016/08/f33/2015-Wind-Technologies-Market-Report-08162016.pdf [accessed 28 January 2017].

DoE. n.d. *Renewable electricity production tax credit (PTC)* [Online]. Washington DC: US Department of Energy. Available: https://energy.gov/savings/renewable-electricity-production-tax-credit-ptc [accessed13 March 2018].

Doh, J. P., Lawton, T. C., and Rajwani, T. 2012. Advancing nonmarket strategy research: Institutional perspectives in a changing world. *Academy of Management Perspectives*, 26, 22–39.

Dow Chemical Company. 2013. Statement for the record to the Senate Energy and Natural Resources Committee hearing on opportunities and challenges for natural gas.

Downey, J. 2013. Duke Energy responds cautiously to Obama's carbon plan. *Charlotte Business Journal*, 26 June 2013.

Downie, C. 2012. Toward an understanding of state behavior in prolonged international negotiations. *International Negotiation*, 17, 295–320.

Downie, C. 2013. Three ways to understand state actors in international negotiations: Climate change in the Clinton years (1993–2000). *Global Environmental Politics*, 13, 22–40.

Downie, C. 2014a. *The politics of climate change negotiations: Strategies and variables in prolonged international negotiations*. Cheltenham: Edward Elgar.

Downie, C. 2014b. Transnational actors in environmental politics: strategies and influence in long negotiations. *Environmental Politics*, 23, 376–394.

Downie, C. 2017a. Business actors, political resistance, and strategies for policymakers. *Energy Policy*, 108, 583–592.

Downie, C. 2017b. Fighting for King Coal's crown: Business actors in the US coal and utility industries. *Global Environmental Politics*, 17, 21–39.

Downie, C. and Drahos, P. 2015. US institutional pathways to clean coal and shale gas: Lessons for China. *Climate Policy*, 17, 246–260.

Drahos, P. 2002. *Information feudalism: Who owns the knowledge economy?*. New York: The New Press.

Duke Energy. 2015. Duke Energy CEO Lynn Good comments on EPA's Clean Power Plan. News release, 3 August 2015. Available: www.duke-energy.com/news/releases/2015080301.asp [accessed 9 June 2015].

Dunbar, Elizabeth. 2014. Minn. could have to reduce carbon by 40% under new rule. *MPR News*, 2 June 2014.

Duncan, M. 2013. ACCCE President Mike Duncan speech to ACCCE Board: Coal-based electricity must endure for generations to come. *Business Wire*, 26 June 2013. Available: www.businesswire.com/news/home/20130626006178/en/Coal-based-Electricity-Endure-Generations#.Ve5LJ1WqpBd

Durbin, M. J. 2015. Testimony of Martin J. Durbin, President and CEO, America's Natural Gas Alliance, US Senate Committee on Energy and Natural Resources.

E&E News. 2016. *The fate of the Obama administration's signature climate change rule is in the hands of the courts* [Online]. E&E Publishing. Available: www.eenews. net/interactive/clean_power_plan/fact_sheets/legal [accessed 25 August 2016].

Ebinger, C. K. and Avasarala, G. 2013. *The case for US liquefied natural gas exports.* Brookings Institute. Available: www.brookings.edu/articles/the-case-for-u-s-liquefied-natural-gas-exports/ [accessed 4 June 2014].

EEI. 2014. *Stock performance: Q4 2014 financial update.* Quarterly report of the US shareholder-owned electric utility industry. Edison Electric Institute. Available: www.eei.org/resourcesandmedia/industrydataanalysis/industryfinancialanalysis/QtrlyFinancialUpdates/Documents/QFU_Stock/2014_Q4_Stock_Performance.pdf [accessed 9 June 2015].

EEI. 2015. EEI Statement on EPA's Clean Power Plan. Press release, August 3, 2015. Available: www.eei.org/resourcesandmedia/newsroom/Pages/Press%20Releases/EEI%20Statement%20on%20EPA%E2%80%99s%20Clean%20Power%20Plan.aspx [accessed 9 June 2015].

Egan, P. J. and Mullin, M. 2017. Climate change: US public opinion. *Annual Review of Political Science*, 20, 209–227.

EIA. 2011. *Global natural gas prices vary considerably* [Online]. Washington DC: US Energy Information Administration. Available: www.eia.gov/todayinenergy/detail.cfm?id=3310# [accessed 25 August 2016].

EIA. 2014. *Major US coal producers, 2014* [Online]. Washington DC: Energy Information Administration. Available: www.eia.gov/coal/annual/pdf/table10.pdf [accessed 12 August 2016].

EIA. 2015a. *Annual coal report 2013.* Washington DC: Energy Information Administration.

EIA. 2015b. *Effects of removing restrictions on US crude oil exports* [Online]. US Energy Information Administration. Available: www.eia.gov/analysis/requests/crude-exports/ [accessed 25 August 2016].

EIA. 2015c. *Renewables share of US energy consumption highest since 1930s* [Online]. Washington DC: US Energy Information Administration. Available: www.eia.gov/todayinenergy/detail.php?id=21412 [accessed 12 June 2016].

EIA. 2016a. *Annual coal report 2015.* Washington DC: Energy Information Administration.

EIA. 2016b. *Coal made up more than 80% of retired electricity generating capacity in 2015* [Online]. Washington DC: US Energy Information Administration. Available: www.eia.gov/todayinenergy/detail.php?id=25272 [accessed 1 May 2017].

EIA. 2016c. *Growth in domestic natural gas production leads to development of LNG export terminals* [Online]. Washington DC: US Enegy Information Administration. Available: www.eia.gov/todayinenergy/detail.cfm?id=25232 [accessed 25 August 2016].

EIA. 2016d. *Henry Hub natural gas spot price* [Online]. Washington DC: US Energy Information Administration. Available: www.eia.gov/dnav/ng/hist/rngwhhdm.htm [accessed 25 August 2016].

EIA. 2016e. International Energy Outlook 2016. Washington DC: Energy Information Administration.

EIA. 2016f. *What is the role of coal in the United States?* [Online]. Energy Information Administration. Available: www.eia.gov/energy_in_brief/article/role_coal_us.cfm [accessed 30 October 2016].

EIA. 2016g. *Wind adds the most electric generation capacity in 2015, followed by natural gas and solar* [Online]. Washington DC: US Energy Information Administration. Available: www.eia.gov/todayinenergy/detail.php?id=25492 [accessed 14 June 2017].

EIA. 2016h. *Wind generation share exceeded 10% in 11 states in 2015* [Online]. Washington DC: US Energy Information Administration. Available: www.eia.gov/todayinenergy/detail.php?id=28512 [accessed 1 May 2017].

EIA. 2017a. *Annual coal report.* Washington DC: United States Energy Information Administration.

EIA. 2017b. *Annual energy outlook 2017.* Washington DC: Energy Information Administration

EIA. 2017c. *Most US nuclear power plants were built between 1970 and 1990* [Online]. Washington DC: US Energy Information Administration. Available: www.eia.gov/todayinenergy/detail.php?id=30972 [accessed 12 December 2017].

EIA. 2017d. *US energy-related carbon dioxide emissions, 2016.* Washington DC: Energy Information Administration.

EIA. 2017e. *US energy-related carbon dioxide emissions, 2016.* Washington DC: United States Energy Information Administration.

EIA. 2017f. *United States remains the world's top producer of petroleum and natural gas hydrocarbons* [Online]. Washington DC: US Energy Information Administration. Available: www.eia.gov/todayinenergy/detail.php?id=31532 [accessed 1 June 2017].

EIA. 2017g. *Utility-scale solar has grown rapidly over the past five years* [Online]. Washington DC: US Energy Information Administration. Available: www.eia.gov/todayinenergy/detail.php?id=31072 [accessed 1 June 2017].

EIA. 2017h. *Wind turbines provide 8% of US generating capacity, more than any other renewable source* [Online]. Washington DC: US Energy Information Administration. Available: www.eia.gov/todayinenergy/detail.php?id=31032 [accessed 1 June 2017].

EIA. 2018a. *Annual energy outlook 2018.* Washington DC: United States Energy Information Administration.

EIA. 2018b. *By some measures, US natural gas production set a record in 2017* [Online]. United States Energy Information Administration. Available: www.eia.gov/todayinenergy/detail.php?id=35712 [accessed 3 June 2017].

EIA. 2018c. *EIA forecasts natural gas to remain primary energy source for electricity generation* [Online]. United States Energy Information Administration. Available: www.eia.gov/todayinenergy/detail.php?id=34612 [accessed 3 June 2017].

EIA. 2018d. *EIA projects that US coal demand will remain flat for several decades* [Online]. United States Energy Information Administration. Available: www.eia.gov/todayinenergy/detail.php?id=35572 [accessed 3 June 2017].

EIA. 2018e. *Electricity in the United States* [Online]. United States Energy Information Administration. Available: www.eia.gov/energyexplained/index.php?page=electricity_in_the_united_states [accessed 14 September 2018].

EIA. 2018f. *US energy-related CO2 emissions expected to rise slightly in 2018, remain flat in 2019* [Online]. United States Energy Information Administration. Available: www.eia.gov/todayinenergy/detail.php?id=34872 [accessed 3 June 2018].

EIA. 2018g. *United States remains the world's top producer of petroleum and natural gas hydrocarbons* [Online]. United States Energy Information Administration. Available: www.eia.gov/todayinenergy/detail.php?id=36292&src=email [accessed 14 September 2018].

EIA. 2018h. *US energy facts* [Online]. United States Energy Information Administration. Available: www.eia.gov/energyexplained/?page=us_energy_home [accessed 3 June 2017].

Eisenberg, R. 2013a. Comments of the National Association of Manufacturers on the 2012 LNG Export Study. www.fossil.energy.gov/programs/gasregulation/authorizations/export_study/ross_eisenberg_em01_24_13.pdf: National Association of Manufacturers [accessed 14 June 2017].

Eisenberg, R. 2013b. Testimony of Ross Eisenberg, Vice President, energy and resources policy, National Association of Manufacturers, before the Senate Committee on Energy and Natural Resources on 'Opportunity and challenges associated with America's natural gas resources'.

Elsner, G. 2014. Koch network, fossil-fuel front groups lobby congress against wind-energy tax breaks. *Huff Post* [Online]. 19 August. Available: www.huffingtonpost.com/gabe-elsner/koch-network-fossil-fuel-_b_5509075.html.

EPA. 2009a. *Endangerment and cause or contribute findings for greenhouse gases under section 202(a) of the Clean Air Act* [Online]. Available: www.epa.gov/climatechange/endangerment/ [accessed 7 September 2015].

EPA. 2009b. *A preliminary analysis of the Waxman–Markey discussion draft, the American Clean Energy and Security Act of 2009 in the 111th Congress*. Washington DC: Environmental Protection Agency.

EPA. 2013. Proposed carbon pollution standards for new power plants – September 20, 2013. United States Environmental Protection Agency. Available: www2.epa.gov/carbon-pollution-standards/2013-proposed-carbon-pollution-standard-new-power-plants [accessed 1 May 2016].

EPA. 2014. *Comments of Southern Company*. Washington DC: Environmental Protection Agency. Washington DC: www.regulations.gov/index.jsp#!documentDetail;D=EPA-HQ-OAR-2013-0602-22907 [accessed 14 June 2014].

EPA. 2015. Clean Power Plan final rule. *Environmental Protection Agency*, 3 August 2015.

Falkner, R. 2008. *Business power and conflict in international environmental politics*. New York: Palgrave Macmillan.

Federal Information & News Dispatch, Inc. 2015. Alexander, six senators oppose wasting up to $9.4 billion on extension of wind production tax credit. News release.

Feldscher, K. 2015. Kyle Feldscher. *Washington Examiner*, 16 September.

Fifield, A. 2013. Siemens calls for long-haul US wind strategy. *Financial Times*.

Figueiredo, J. M. and Richter, B. K. 2014. Advancing the empirical research on lobbying. *Annual Review of Political Science*, 17, 163–185.

Figueiredo, J. M. and Tiller, E. H. 2001. The structure and conduct of corporate lobbying: How firms lobby the Federal Communications Commission. *Journal of Economics & Management Strategy*, 10, 91–122.

Figueres, C., Schellnhuber, H. J., Whiteman, G., Rockström, J., Hobley, A., and Rahmstorf, S. 2017. Three years to safeguard our climate. Vol. 546. Available: www.nature.com/news/three-years-to-safeguard-our-climate-1.22201 [accessed 9 November 2017].

Fisher-Vanden, K. 2000. International policy instrument prominence in the climate change debate. *In:* Harris, P. (ed.) *Climate change and American foreign policy*. New York: St Martin's Press.

Friedman, L. 2017. Trump administration moves to open Arctic refuge to drilling studies. *New York Times*.

Furlong, S. R. and Kerwin, C. M. 2005. Interest group participation in rule making: A decade of change. *Journal of Public Administration Research and Theory: J-PART*, 15, 353–370.

Galik, C. S., Decarolis, J. F., and Fell, H. 2017. Evaluating the US mid-century strategy for deep decarbonization amidst early century uncertainty. *Climate Policy*, 17, 1046–1056.

GAO. 2014. *Changing crude oil markets: Allowing exports could reduce consumer fuel prices, and the size of the strategic reserves should be reexamined.* United States Government Accountability Office. Available: www.gao.gov/assets/670/666 274.pdf [accessed 19 November 2014].

Garland, M. and Reilly, S. 2015. Setting the record straight on the wind energy tax credit. *The Hill* [Online]. Available: https://thehill.com/blogs/congress-blog/ energy-environment/244569-setting-the-record-straight-on-the-wind-energy-tax [accessed 11 June 2015].

Gebrekidan, S. 2014. Americans choose savings at the pump over oil exports: Reuters/Ipsos poll. *Reuters*, 20 March.

Geman, B. 2013. Manufacturers go to war with oil industry over gas exports. *The Hill*, 10 January.

Gerard, J. N. 2013. Written testimony of Jack N. Gerard, President and CEO of American Petroleum Institute, hearing on 'Opportunities and challenges for natural gas' before the US Senate Committee on Energy and Natural Resources.

Gerard, J. N. 2014. *Jack Gerard delivers state of American energy address* [Online]. American Petroleum Institute. Available: www.api.org/news-policy-and-issues/ news/2018/01/09/apis-jack-gerard-delivers-the-2018-state-of-american-energy-speech [accessed 11 January 2018].

Gerard, J. N. 2015. *2015 state of American energy.* Washington DC: American Petrolem Institute.

Gerring, J. 2017. Qualitative methods. *Annual Review of Political Science*, 20, 15–36.

Gilbert, K. 2014. Could a GOP Congress back renewable energy? *Institutional Investor*, 24 December 2014.

Gill, S. R. and Law, D. 1989. Global hegemony and the structural power of capital. *International Studies Quarterly*, 33, 475–499.

Goggin, M. 2014. *The facts about wind energy's impacts on electricity markets: Cutting through Exelon's claims about 'negative prices' and 'market distortion'.* American Wind Energy Association. Available: www.awea.org/Awea/media/ Resources/Publications%20and%20Reports/White%20Papers/AWEA-white-paper-Cutting-through-Exelon-s-claims.pdf

Goldenberg, S. 2014. Conservative groups bid to wreck Obama's proposal on carbon emissions. *Guardian*, 31 May 2014.

Goode, D. 2014. Democrats divided on supplying more. *Politico*. 23 July.

GreenBiz. 2009. Duke leaves coal group over anti-climate stance. *GreenBiz*, 3 September 2009.

Greenblatt, J. B. and Wei, M. 2016. Assessment of the climate commitments and additional mitigation policies of the United States. *Nature Clim. Change*, 6, 1090–1093.

Gross, D. 2015. It's a wonderful life for the solar industry right now. *Grist*, December 25, 2015.

Grossman, G. and Helpman, E. 2001. *Special interest politics.* Cambridge: The MIT Press.

Guardian. 2015. Iran deal 'adoption day': US approves conditional sanctions waivers. *Guardian*, 19 October 2015.

Guber, D. and Bosso, C. 2007. Framing ANWR: Citizens, consumers and the privileged position of business. *In:* Kraft, M. and Kamieniecki, S. (eds.) *Business and environmental policy: Corporate interests in the American political system.* Cambridge: MIT Press.

Gunningham, N., Kagan, R. A., and Thornton, D. 2004. Social license and environmental protection: Why businesses go beyond compliance. *Law & Social Inquiry*, 29, 307–341.

Hadden, J. 2015. *Networks in contention.* Cambridge: Cambridge University Press.

Håkansson, H. and Ford, D. 2002. How should companies interact in business networks? *Journal of Business Research*, 55, 133–139.

Hall, P. A. and Thelen, K. 2009. Institutional change in varieties of capitalism. *Socio-Economic Review*, 7, 7–34.

Hall, P. and Soskice, D. 2001. An introduction to varieties of capitalism. *In:* Hall, P. and Soskice, D. (eds.) *Varieties of capitalism: The institutional foundations of comparative advantage.* Oxford: Oxford University Press.

Halperin, D. 2014. US gas exports flow through DC lobbying's revolving door. *The Huffington Post*, 26 June 2014.

Haq, A. 2015. ALEC's latest attack on the Clean Power Plan is fizzling. *Business Spectator*, 20 February 2015.

Harbert, K. 2013. Statement of Karen Harbet during EPA listening session. Institute for 21st Century Energy, US Chamber of Commerce.

Harder, A. 2014. Oil producers to pump up lobbying to remove US export ban. *The Wall Street Journal*, 24 August 2014.

Harder, A. and Berthelsen, C. 2015. End of oil-export ban provides blueprint for bipartisan compromise. *The Wall Street Journal*, 20 December 2015.

Harrison, K. and Sundstrom, L. 2007. The comparative politics of climate change. *Global Environmental Politics*, 7, 1–18.

Heede, R. 2014. Tracing anthropogenic carbon dioxide and methane emissions to fossil fuel and cement producers, 1854–2010. *Climatic Change*, 122, 229–241.

Heitkamp, H. 2015. *Heitkamp offers amendments to keystone bill to advance key national energy policies.* Washington DC: United States Senate.

Hervé, A. 2014. Roles of brokerage networks in transnational advocacy networks. *Environmental Politics*, 23, 395–416.

Hess, D. J. 2013. Industrial fields and countervailing power: The transformation of distributed solar energy in the United States. *Global Environmental Change*, 23, 847–855.

Hess, D. J. 2014. Sustainability transitions: A political coalition perspective. *Research Policy*, 43, 278–283.

Hillman, A. J., Keim, G. D., and Schuler, D. 2004. Corporate political activity: A review and research agenda. *Journal of Management*, 30, 837–857.

Hirsh, R. 1999. *Power loss: The origins of deregulation and restructuring in the american electric utility system.* Cambridge: MIT Press.

Holland, S. P., Hughes, J. E., Knittel, C. R., and Parker, N. C. 2014. Some inconvenient truths about climate change policy: The distributional impacts of transportation policies. *Review of Economics and Statistics*, 97, 1052–1069.

Hughes, J. 2016. Wholesale price of solar coming down: First Solar CEO. Available: www.msn.com/en-us/money/video/wholesale-price-of-solar-coming-down-first-solar-ceo/vi-AAhrc5h [accessed 17 June 2017].

Hughes, L. and Urpelainen, J. 2015. Interests, institutions, and climate policy: Explaining the choice of policy instruments for the energy sector. *Environmental Science & Policy*, 54, 52–63.

Hula, K. 1999. *Lobbying together: Interest group coalitions in legislative politics.* Washington DC: Georgetown University Press.

Hulse, C. and Herszenhorn, D. M. 2010. Democrats call off climate bill effort. *New York Times*, 22 July 2010.

IEA. 2008. *World energy outlook 2008*. Paris: International Energy Agency.

IEA. 2014a. *Energy policies of IEA countries: The United States 2014 review.* Paris: International Energy Agency.

IEA. 2015a. *Energy and climate change*. Paris: International Energy Agency.

IEA. 2015b. *World energy outlook 2015*. Paris: International Energy Agency.

IEA. 2016a. *Key world energy statistics*. Paris: International Energy Agency.

IEA. 2016b. *World energy outlook 2016*. Paris: International Energy Agency.

IEA. 2017a. *Key world energy statistics*. Paris: International Energy Agency.

IEA. 2017b. *Tracking progress: Renewable power* [Online]. International Energy Agency. Available: www.iea.org/etp/tracking2017/renewablepower/ [accessed 2 February 2017].

IEA. 2017c. *World energy outlook 2017*. Paris: International Energy Agency.

IEA. 29 September 2014b. How solar energy could be the largest source of electricity by mid-century. Available: www.iea.org/newsroomandevents/pressreleases/2014/september/how-solar-energy-could-be-the-largest-source-of-electricity-by-mid-century.html [accessed 1 May 2016].

IEA/IRENA. 2017. *Perspectives for the energy transition – investment needs for a low-carbon energy system.* Paris: IEA and IRENA.

Institute for 21st Century Energy. 2015. *Renewables* [Online]. US Chamber of Commerce. Available: www.energyxxi.org/renewables [accessed 7 September 2015].

IPAA. 2013. *Profile of independent producers 2012–2013.* Washington DC: Independent Petroleum Association of America.

IPAA. 2014. IPAA responds to claims agains crude oil exports. Independent Petroleum Association of America. Available: www.ipaa.org/ipaa-responds-claims-crude-oil-exports/ [accessed 9 August 2014].

IPCC. 2014. Summary for policymakers. *In*: Field, C. B. B., Barros, V. R., Dokken, D. J., Mach, K. J., Mastrandrea, M. D., Bilir, T. E., Chatterjee, M., Ebi, K. L., Estrada, Y. O., Genova, R. C., Girma, B., Kissel, E. S., Levy, A. N., MacCracken, S., Mastrandrea, P. R., and White, L. L. (eds) *Climate change 2014: Impacts, adaptation, and vulnerability. Part A: Global and sectoral aspects. Contribution of working group II to the fifth assessment report of the intergovernmental panel on climate change.* Cambridge, UK, and New York, USA: Cambridge University Press. Available: www.ipcc.ch/pdf/assessment-report/ar5/wg2/ar5_wgII_spm_en.pdf

Jacobsson, S. and Lauber, V. 2006. The politics and policy of energy system transformation—explaining the German diffusion of renewable energy technology. *Energy Policy*, 34, 256–276.

Jett, J. 2014. Last man standing: Bob Murray and the war on coal. *West Virginia Executive*, 30 May 2014.

Joachim, J. 2003. Framing issues and seizing opportunities: The UN, NGOs, and women's rights. *International Studies Quarterly*, 47, 247–274.

Jones, C. A. and Levy, D. L. 2007. North American business strategies towards climate change. *European Management Journal*, 25, 428–440.

Juliano, N. 2014. Politically wired nonprofit hooks up us industry with countries seeking LNG. *E&E News*, 13 November 2014.

Kagan, R. A. 1991. Adversarial legalism and American government. *Journal of Policy Analysis and Management*, 10, 369–406.

Kagan, R. A. 2007. Globalization and legal change: The 'Americanization' of European law? *Regulation & Governance*, 1, 99–120.

Kary, T., Loh, T., and Polson, J. 2016. Coal slump sends mining giant Peabody Energy into bankruptcy. *Bloomberg*, 13 April.

Kauffman, C. 2017. *Grassroots global governance: Local watershed management experiments and the evolution of sustainable development.* Oxford: Oxford University Press.

Kelsey, S. 2014. *The green spiral: Policy-industry feedback and the success of international environmental negotiation.* PhD thesis. Berkeley, CA: University of California.

Keohane, N., Revesz, R., and Stavins, R. 1998. The choice of regulatory instruments in environmental policy. *Harvard Environmental Law Review*, 22, 313–367.

Keohane, R. and Nye, J. (eds.) 1972. *Transnational relations and world politics.* Cambridge: Harvard University Press.

Khan, F. 2013. Testimony to US Senate Committee on Energy and Natural Resources by Faisel Khan, Managing Director, Citi Research. 16 July 2013.

Khedr, O. 2015. *IBISWorld industry report OD4494: Solar panel installation in the US.* Melbourne: IBISWorld.

Kim, S. E., Urpelainen, J., and Yang, J. 2016. Electric utilities and American climate policy: Lobbying by expected winners and losers. *Journal of Public Policy*, 36, 251–275.

Klesse, W. R. 2013. Testimony of William R. Klesse, Chairman of the Board and CEO, Valero Energy Corporation before the US Senate Committee on Energy and Natural Resources.

Klyza, C. M. and Sousa, D. 2008. *American environmental policy, 1990–2006: Beyond gridlock*, Cambridge: The MIT Press.

Knox-Hayes, J. 2012. Negotiating climate legislation: Policy path dependence and coalition stabilization. *Regulation & Governance*, 6, 545–567.

Kollewe, J. and Farrell, S. 2015. Shell agrees to buy BG group for £47bn. *Guardian*, 8 April 2015.

Korosec, K. 2015. In US, there are twice as many solar workers as coal miners. *Fortune*, 16 January 2015.

Kraft, M. and Kamieniecki, S. 2007. Analyzing the role of business in environmental policy. *In:* Kraft, M. and Kamieniecki, S. (eds.) *Business and environmental policy: Corporate interests in the American political system.* Cambridge: MIT Press.

Krauss, C. and Galbraith, K. 2009. Climate bill splits Exelon and US Chamber. *New York Times*, 28 September 2009.

Krauss, C. and Schwartz, N. D. 2013. Foreseeing trouble in exporting natural gas. *New York Times*, 15 August 2013.

Lacey, S. 2015. Congress passes tax credits for solar and wind: 'Sausage-making at its most intense. Available: www.greentechmedia.com/articles/read/breaking-house-passes-1-1-trillion-spending-bill-with-renewable-energy-tax [accessed 18 December 2015].

Lacey, S. 2016. *How the national solar lobby passes the investment tax credit.* 4 January 2016. Available: www.greentechmedia.com/articles/read/inside-the-solar-industrys-successful-bid-to-pass-the-investment-tax-credit [accessed 12 June 2016].

Laird, F. N. and Stefes, C. 2009. The diverging paths of German and United States policies for renewable energy: Sources of difference. *Energy Policy*, 37, 2619–2629.

Lance, R. 2015. Testimony of Ryan Lance, before the US Committee on Energy and Natural Resources, United States Senate Hearing on US Crude Oil Export Policy.

Landler, M. 2014. US and China reach climate accord after months of talks. *New York Times*. Available: www.nytimes.com/2014/11/12/world/asia/china-us-xi-obama-apec.html [accessed 14 November 2014].

Lavelle, M. and Donald, D. 2009. Southern Company dominates the climate lobbying scene. The Center for Public Integrity.

Layzer, J. 2007. Deep Freeze: How business has shaped the global warming debate in congress. *In:* Kraft, M. and Kamieniecki, S. (eds.) *Business and environmental policy: Corporate interests in the American political system.* Cambridge: MIT Press.

Layzer, J. 2012. *Open for business: Conservatives' opposition to environmental regulation.* Cambridge: MIT Press.

Leiserowitz, A., Maibach, E., and Roser-Renouf, C. 2008. Climate change in the American mind: American's climate change beliefs, attitudes, policy preferences, and actions. Yale Project on Climate Change and George Mason University Center for Climate Change Communication.

Lelong, C., Currie, J., Dart, S., and Koenig, P. 2013. *The window for thermal coal investment is closing.* New York: Goldman Sachs.

Lessig, L. 2011. *Republic lost: How money corrupts Congress – and a plan to stop it.* New York: Twelve.

Levin, K., Cashore, B., Bernstein, S., and Auld, G. 2012. Overcoming the tragedy of super wicked problems: Constraining our future selves to ameliorate global climate change. *Policy Sciences*, 45, 123–152.

Levy, D. and Kolk, A. 2002. Strategic responses to global climate change: Conflicting pressures on multinationals in the oil industry. *Business and Politics*, 4, 275–300.

Levy, D. L. and Newell, P. J. 2002. Business strategy and international environmental governance: Toward a neo-Gramscian synthesis. *Global Environmental Politics*, 2, 84–101.

Lindblom, C. 1977. *Politics and markets: The world's political economic system.* New York: Basic Books.

Lipton, E. 2014. Energy firms in secretive alliance with attorneys general. *New York Times* [accessed 2 February 2015].

Lipton, E. and Krauss, C. 2015. Oil industry gaining push for repeal of US ban on petroleum exports. *New York Times*, 5 October 2015.

Lizza, R. 2010. As the world burns. *The New Yorker*, 11 October 2010.

Luke, S. 1974. *Power: A radical view.* Houndmills: MacMillan Education.

Maher, K. 2013. Big coal to fight Obama plan. *The Wall Street Journal*, 26 June 2013.

Mahoney, C. 2007. Networking vs. allying: The decision of interest groups to join coalitions in the US and the EU. *Journal of European Public Policy*, 14, 366–383.

Markey, E. 2013. Markey stands with American people and manufacturers on natural gas exporting. 10 January. Available: www.markey.senate.gov/news/press-releases/markey-stands-with-american-people-and-manufacturers-on-natural-gas-exporting [accessed 2 March 2014].

McAdam, D. 2017. Social movement theory and the prospects for climate change activism in the United States. *Annual Review of Political Science*, 20, 189–208.

McDonnell, T. 2015. Coal-country states declare war on Obama's climate rules. *Grist*, 20 February 2015.

McFarland, A. S. 1984. Energy lobbies. *Annual Review of Energy*, 9, 501–527.

McGlade, C. and Ekins, P. 2015. The geographical distribution of fossil fuels unused when limiting global warming to 2°C. *Nature*, 517, 187–190.

McGuiness, K. 2015. Renewable industry wraps up a great 2014, but future of tax incentives is murky. 7 January 2015. American Council on Renewable Energy [Online]. Available: www.acore.org/acore-blog/item/4200-renewable-industry-wraps-up-a-huge-2014-but-future-of-tax-incentives-is-murky [accessed 2 June 2015].

McTague, J. 2014. Solar's power is limited in Congress. *Barron's Asia*, 21 June 2014.

Meckling, J. 2011. *Carbon coalitions: Business, climate politics, and the rise of emissions trading*. Cambridge: The MIT Press.

Meckling, J. 2015. Oppose, support, or hedge? distributional effects, regulatory pressure, and business strategy in environmental politics. *Global Environmental Politics*, 15, 19–37.

Meckling, J. and Hughes, L. 2015. Globalizing solar: Industry specialization and firm demands for trade protection. *Berkeley Roundtable on the International Economy*. (BRIE) Working Paper 2015-5. Available: SSRN: https://ssrn.com/abstract=2638833 or http://dx.doi.org/10.2139/ssrn.2638833

Meckling, J., Kelsey, N., Biber, E., and Zysman, J. 2015. Winning coalitions for climate policy. *Science*, 349, 1170–1171.

Mellahi, K., Frynas, J. G., Sun, P., and Siegel, D. 2016. A review of the nonmarket strategy literature. *Journal of Management*, 42, 143–173.

Meyer, D. S. 2004. Protest and political opportunities. *Annual Review of Sociology*, 30, 124–145.

Meyer, D. S. and Minkoff, D. C. 2004. Conceptualizing political opportunity. *Social Forces*, 82, 1457–1492.

Meyers, R. 2014. Marathon Petroleum warns against lifting crude export ban as profits triple. *FuelFix*. Available: https://fuelfix.com/blog/2014/10/30/marathon-petroleums-profits-triple/ [accessed 2 February 2015].

Mikulka, J. 2015. Lifting ban on US crude oil export would enable massive fracking expansion. 10 August 2015. *Desmog* [Online]. Available: www.desmogblog.com/2018/10/15/us-oil-exports-exceeding-predictions-fracking-gas-prices-china [accessed 18 October 2018].

Mildenberger, M. 2013. The politics of strategic accommodation: Explaining business support for US climate policy. *International Conference on Public Policy*. Grenoble: June.

Miliband, R. 1969. *The state in capitalist society*. London: Weidenfeld and Nicolson.

Ministry of Foreign Affairs Japan. 2013. *Japan–US summit meeting* [Online]. Available: www.mofa.go.jp/region/n-america/us/pmv_1302/130222_01.html [accessed 25 August 2016].

MIT. 2015. *The future of solar energy*. Cambridge, MA: Massachusetts Institute of Technology.

Montgomery, L. 2014. Inside the collapse of the year's biggest tax deal. *Washington Post*, 11 December.

Morford, S. 2009. Exelon latest to leave US Chamber of Commerce; Is Nike next?. *InsideClimate News*. [accessed 2 June 2014].

Morgan, K. and Orloff, A. S. 2017. Introduction. *In:* Morgan, K. and Orloff, A. S. (eds.) *The many hands of the state*. Cambridge: Cambridge University Press.

Mufson, S. 2015. Shell Oil will drop its membership in ALEC, citing differences over climate change. *Washington Post*, 7 August 2015.

Mügge, D. 2011. Limits of legitimacy and the primacy of politics in financial governance. *Review of International Political Economy*, 18, 52–74.

Mundy, A. 2014a. Lift the crude-oil export ban, US Chamber says. *The Wall Street Journal*, 8 January 2014.

Mundy, A. 2014b. Refiners form coaltion to fight exports of crude oil. *The Wall Street Journal*, 12 March 2014.

NAM. 9 September 2015. *RE: Tax extenders letter*. To United States Congress. Available: www.nam.org/Issues/Tax-and-Budget/Tax-Extenders-Letter-09-10-2015.pdf

NASA. 2017. *NASA, NOAA data show 2016 warmest year on record globally*. Washington: National Aeronautics and Space Administration.

NERA. 2012. *Macroeconomic impacts of LNG exports from the United States*. Washington DC: NERA Economic Consulting.

NERA. 2014. Potential energy impacts of the epa proposed clean power plan. NERA Economic Consulting. Available: www.nera.com/publications/archive/2014/potential-impacts-of-the-epa-clean-power-plan.html [accessed 2 June 2016].

Newell, P. 2000. *Climate for change: Non-state actors and the global politics of the greenhouse*. Cambridge: Cambridge University Press.

Newell, P. and Paterson, M. 1998. A climate for business: Global warming, the state and capital. *Review of International Political Economy*, 5, 679–703.

NPR. 2015. Murkowski critical of proposal for Arctic national wildlife refuge. National Public Radio.

O'Neill, K., Weinthal, E., Suiseeya, K. R. M., Bernstein, S., Cohn, A., Stone, M. W., and Cashore, B. 2013. Methods and global environmental governance. *Annual Review of Environment and Resources*, 38, 441–471.

Odell, J. and Sell, S. 2006. Reframing the issue: The WTO coalition on intellectual property and public health, 2001. *In:* Odell, J. (ed.) *Negotiating trade: Developing countries in the WTO and NAFTA*. Cambridge: Cambridge University Press.

OECD. 2015. *OECD companion to the inventory of support measures for fossil fuels 2015*. Paris: OECD Publishing.

Office of Management and Budget. 2015. Fiscal year 2016: Budget of the US government. Available: www.gpo.gov/fdsys/pkg/BUDGET-2016-APP/pdf/BUDGET-2016-APP.pdf [accessed 12 June 2017].

Osten, M. 2015. *IBISWorld industry report 33441c: Solar panel manufacturing in the US*. Melbourne: IBISWorld.

Parkinson, G. 2012. Interview: First Solar CEO James Hughes. Available: http://reneweconomy.com.au/2012/interview-first-solar-ceo-james-hughes-72086 [accessed 9 August 2014].

Parra, F. 2004. *Oil politics: A modern history of petroleum*. London, I.B. Tauris.

PBEF. 2015. *Reliable and affordable energy* [Online]. Partnership for a Better Energy Future. Available: www.betterenergyfuture.org/ [accessed 7 September 2015].

Peabody Energy. 2014. *EPA carbon rules: Why Americans should engage* [Online]. www.peabodyenergy.com/content/494/coal-in-the-united-states/epa-carbon-rules. Available: www.peabodyenergy.com/content/494/coal-in-the-united-states/epa-carbon-rules [accessed 25 August 2016].

Pegels, A. and Lütkenhorst, W. 2014. Is Germany's energy transition a case of successful green industrial policy? Contrasting wind and solar PV. *Energy Policy*, 74, 522–534.

Pew Center on Global Climate Change. 2009. At a glance: American Clean Energy and Security Act of 2009. Available: www.c2es.org/site/assets/uploads/2009/06/Waxman-Markey-short-summary-revised-June26.pdf [accessed 12 June 2014].

Pfeiffer, A., Millar, R., Hepburn, C., and Beinhocker, E. 2016. The '2°C capital stock' for electricity generation: Committed cumulative carbon emissions from the electricity generation sector and the transition to a green economy. *Applied Energy*, 179, 1395–1408.

Pierson, P. 2004. *Politics in time: History, institutions, and social analysis*. Princeton, NJ: Princeton University Press.

Plautz, J. 2013. Unlikely firms bring clout and cash to clean energy lobbying effort. *Inside Climate News*, 22 January.

Podolny, J. M. and Page, K. L. 1998. Network forms of organization. *Annual Review of Sociology*, 24, 57–76.

Pooley, E. 2008. McCain's gift to the green movement. *Time Magazine*, 14 June 2014.

Popovich, N. and Albeck-Ripka, L. 2017. *52 Environmental rules on the way out under Trump* [Online]. 6 October 2017. *New York Times*. Available: www.nytimes.com/interactive/2017/10/05/climate/trump-environment-rules-reversed.html [accessed 2 February 2018].

Prakash, A. 2000. *Greening the firm: The politics of corporate environmentalism*. Cambridge: Cambridge University Press.

Przeworski, A. and Teune, H. 1970. *The logic of comparative social inquiry*. New York: Wiley-Interscience.

Putnam, R. 1988. Diplomacy and domestic politics: The logic of two-level games. *International Organization*, 42, 427–460.

Quiñones, M. 2014. Mining group rolls out ads in key states ahead of EPA climate rule. *E&E News*, 20 May 2014.

Raustiala, K. 1997. States, NGOs, and international environmental institutions. *International Studies Quarterly*, 41, 719–740.

REN21. 2015. *Renewables 2015: Global status report*. Paris: Renewable Energy Policy Network for the 21st Century.

Resch, R. 2014. Rhone Resch opening remarks at SPI 2014. *Solar Energy Industries Association*. Available: www.seia.org/news/rhone-resch-opening-remarks-spi-2014 [accessed 2 June 2014].

Reuters. 2015. House passes bill axing oil export ban, Veto looms. 9 October 2015.

Rhodes, R. 2006. Policy network analysis. *In:* Moran, M., Rein, M., and Goodin, R. (eds.) *The Oxford handbook of public policy*. Oxford: Oxford University Press.

Rio Tinto. 2009. *Proposed emissions trading legislation in Australia* [Online]. Available: www.riotinto.com/ourcommitment/features-2932_3267.aspx [accessed 9 June 2016].

Rio Tinto. 2010. *Annual report* [Online]. Available: www.riotinto.com/documents/Investors/RioTinto_Annual_report_2010.pdf [accessed 9 June 2016].

Risse-Kappen, T. 1995. Bringing transnational relations back in: introduction. *In:* Risse-Kappen, T. (ed.) *Bringing transnational relations back in: Non-state actors, domestic structures and international institutions*. Cambridge: Cambridge University Press.

Rogelj, J., Den Elzen, M., Höhne, N., Fransen, T., Fekete, H., Winkler, H., Schaeffer, R., Sha, F., Riahi, K., and Meinshausen, M. 2016. Paris Agreement climate proposals need a boost to keep warming well below 2°C. *Nature*, 534, 631–639.

Roselund, C. 2016. Reports of SunEdision in debter-in-possession talks, danger of bankruptcy. *PV Magazine*, 22 March 2016.

Rowell, A. 2014. Big oil sets up crude export lobby group. 27 October 2014. Available: http://priceofoil.org/2014/10/27/big-oil-sets-up-crude-export-lobby-group/ [accessed 12 June 2015].

Ryan, P. 2014. *In response to Russian aggression, key Central European nations plead for US natural gas exports* [Online]. Available: www.speaker.gov/press-release/response-russian-aggression-key-central-european-nations-plead-us-natural-gas-exports [accessed 25 August 2016].

Sabatier, P. 1988. An advocacy coalition framework of policy change and the role of policy-oriented learning therein. *Policy Sciences*, 21, 129–168.

Sampson, A. 1975. *The seven sisters: The great oil companies and the world they shaped*. New York: Viking Press.

Satchwell, A., Mills, A., and Barbose, G. 2015. Quantifying the financial impacts of net-metered PV on utilities and ratepayers. *Energy Policy*, 80, 133–144.

Schattschneider, E. E. 1960. *The semisoveriegn people: A realist's view of democracy in America*. New York: Holt, Rinehart and Winston.

Scheid, B. 2013. Anti-PTC groups dispute job claims. *Inside Energy/Extra*, 7 March.

Schlichting, I. 2013. Strategic framing of climate change by industry actors: A meta-analysis. *Environmental Communication*, 7, 493–511.

Schmidt, V. A. 2010. Taking ideas and discourse seriously: Explaining change through discursive institutionalism as the fourth 'new institutionalism'. *European Political Science Review*, 2, 1–25.

Schultz, F., Kleinnijenhuis, J., Oegema, D., Utz, S., and Van Atteveldt, W. 2012. Strategic framing in the BP crisis: A semantic network analysis of associative frames. *Public Relations Review*, 38, 97–107.

SEIA. 2015a. CQ roll call: Expiring solar tax credite to face a 'slugfest' in Congress. Solar Energy Industries Association [Online]. Available: www.seia.org/news/cq-roll-call-expiring-solar-tax-credit-face-slugfest-congress [accessed 4 October 2016].

SEIA. 2015b. Polling data: Support for solar. Solar Energy Industries Association [Online]. Available: www.seia.org/research-resources/polling-data-support-solar [accessed 7 Sepetmber 2015].

SEIA. 2015c. *Solar investment tax credit (ITC)* [Online]. Solar Energy Industries Association. Available: www.seia.org/policy/finance-tax/solar-investment-tax-credit [accessed 7 September 2015].

Selin, H. and Vandeveer, S. (eds.) 2009. *Changing climates in North American Politics: Institutions, policymaking, and multilevel governance*. Cambridge: MIT Press.

Sell, S. 2003. *Private power, public law: The globalization of intellectual property rights*. Cambridge: Cambridge University Press.

Sell, S. and Prakash, A. 2004. Using ideas strategically: The contest between business and ngo networks in intellectual property rights. *International Studies Quarterly*, 48, 143–175.

Shaffer, B. 1995. Firm-level responses to government regulation: Theoretical and research approaches. *Journal of Management*, 21, 495–514.

Shell. 2015. *Annual report 2015: Selected financial data* [Online]. Available: http://reports.shell.com/annual-report/2015/strategic-report/selected-financial-data.php [accessed 25 August 2016].

Sheppard, K. 2009. *House passes landmark climate and clean-energy bill* [Online]. 27 June. Available: http://grist.org/article/2009-06-26-climate-bill-senate-politics/ full/ [accessed 4 August 2015].

Skjaerseth, J. B. and Skodovin, T. 2003. *Climate change and the oil industry: Common problem, varying strategies.* Manchester: Manchester University Press.

Slaughter, A.-M. 2004. *A new world order.* Princeton, NJ: Princeton University Press.

Smith, C. 2015. Statement of Christopher Smith, Assistant Secretary Office of Fossil Energy, US Department of Energy, before the US Senate Committee on Energy and Natural Resources hearing on S. 33, the LNG Certainty and Transparency Act.

Smith, M. 2000. *American business and political power: Public opinion, elections and democracy.* Chicago, IL: Chicago University Press.

Smith, R. and Bustillo, M. 8 July 2012. Duke chief no stranger to seizing spotlight. *The Wall Street Journal.* Available: www.wsj.com/articles/SB100014240527023 03292204577515003113453474

Snow, D. A. 2007. Framing processes, ideology, and discursive fields. *The Blackwell Companion to Social Movements.* Oxford: Blackwell Publishing Ltd.

Snow, D. A. and Benford, R. D. 1988. Ideology, frame resonance, and participant mobilization. *International Social Movement Research*, 1, 197–217.

SolarCity. 2015. Americans overwhelmingly support extending renewable energy incentives, says national poll. SolarCity. Available: www.solarcity.com/newsroom/ press/americans-overwhelmingly-support-extending-renewable-energy-incentives-says-national [accessed 2 June 2016].

SolarCity. 2015. SolarCity urges swift passage of bipartisan year-end tax package. Available: http://investors.solarcity.com/releasedetail.cfm?releaseid=947176 [accessed 2 June 2016].

SolarWorld. 2015. SolarWorld AG applauds decision in favor of US solar power. Available: www.solarworld.sg/en/the-company/press/current-news/news-detail/ article/solarworld-ag-applauds-decision-in-favor-of-us-solar-power/ [accessed 14 August 2016].

Solomon, B. D. and Krishna, K. 2011. The coming sustainable energy transition: History, strategies, and outlook. *Energy Policy*, 39, 7422–7431.

Southern Company. 2016. *2015 annual report.* Available: http://s2.q4cdn. com/471677839/files/doc_financials/annual2015/2015-annual-report.pdf [accessed 2 June 2016].

Sovacool, B. K. 2009. Rejecting renewables: The socio-technical impediments to renewable electricity in the United States. *Energy Policy*, 37, 4500–4513.

Sovacool, B. K. 2016. How long will it take? Conceptualizing the temporal dynamics of energy transitions. *Energy Research & Social Science*, 13, 202–215.

Stern, N. 2007. *The economics of climate change: The Stern review.* Cambridge: Cambridge University Press.

Stigler, G. 1975. *The theory of economic regulation: Citizen and the state.* Chicago, IL: University Chicago Press.

Stokes, L. C. 2015. *Power politics: Renewable energy policy change in US states.* Cambridge, MA: MIT.

Stokes, L. C. and Breetz, H. L. 2018. Politics in the US energy transition: Case studies of solar, wind, biofuels and electric vehicles policy. *Energy Policy*, 113, 76–86.

Strange, S. 1996. *The retreat of the state: The diffusion of power in the world economy*. Cambridge: Cambridge University Press.

Sweet, C. and Chernova, Y. 2011. First solar revamps amid weak market. *The Wall Street Journal*, 15 December 2011.

Tarrow, S. 1994. *Power in movement*. Cambridge: Cambridge University Press.

Tarrow, S. 2005. *The new transnational activism*. New York: Cambridge University Press.

The Solar Foundation. 2015. National solar jobs census: 2014. The Solar Foundation. Available: www.thesolarfoundation.org/national/

The White House. 2009. *Remarks of President Barack Obama – Address to joint session of Congress*. Washington DC: Office of the Press Secretary. Available: https://obamawhitehouse.archives.gov/the-press-office/remarks-president-barack-obama-address-joint-session-congress [accessed 4 October 2015].

The White House. 2013a. The President's climate action plan. Available: https://obamawhitehouse.archives.gov/the-press-office/2013/06/25/presidential-memorandum-power-sector-carbon-pollution-standards [accessed 4 October 2015].

The White House. 2013b. *Presidential memorandum: Power sector carbon pollution standards*. Washington DC: Office fo the Press Secretary.

The White House. 2015. American businesses act on climate pledge. Available: https://obamawhitehouse.archives.gov/climate-change/pledge

The White House. 2017a. *Presidential executive order on promoting energy independence and economic growth* [Online]. 28 March. Available: www.whitehouse.gov/presidential-actions/presidential-executive-order-promoting-energy-independence-economic-growth/ [accessed 2 February 2018].

The White House. 2017b. *Remarks by President Trump at signing of executive order on an America – first offshore energy strategy* [Online]. 28 April. Available: www.whitehouse.gov/briefings-statements/remarks-president-trump-signing-executive-order-america-first-offshore-energy-strategy/ [accessed 2 February 2018].

The White House. 2017c. *Statement by President Trump on the Paris Climate Accord* [Online]. 1 June. Available: www.whitehouse.gov/the-press-office/2017/06/01/statement-president-trump-paris-climate-accord [accessed 2 February 2018].

Tienhaara, K. 2014. Business: Corporate and industrial influence. *In:* Harris, P. (ed.) *Handbook of global environmental politics*. London: Routledge.

Tienhaara, K., Orsini, A., and Falkner, R. 2012. Global corporations. *In:* Biermann, F. and Pattberg, P. (eds.) *Global Environmental governance reconsidered*. Cambridge: The MIT Press.

Timothy, C. 2015. *Senate panel votes to lift oil export ban* [Online]. 30 July. Available: http://thehill.com/policy/energy-environment/249804-senators-vote-to-lift-oil-export-ban [accessed 2 June 2016].

TOGY. 2015. *CEOs urge lifting of oil export ban* [Online]. Available: www.theoilandgasyear.com/news/ceos-urge-lifting-of-oil-export-ban/ [accessed 12 January 2017].

Tollison, R. D. 2012. The economic theory of rent seeking. *Public Choice*, 152, 73–82.

Trabish, H. K. 2013. Civil war between utility-scale solar and rooftop solar? Available: www.greentechmedia.com/articles/read/civil-war-between-utility-scale-solar-and-rooftop-solar [accessed 2 April 2014].

Trabish, H. K. 2014. Utility Exelon wants to kill wind and solar subsidies while keeping nukes. *Green Tech Media*, 1 April.

Truman, D. 1953. *The governmental process: Political interests and public opinion*. New York, Alfred A. Knopf.

Trumbore, P. 1998. Public Opinion as a domestic constraint in international negotiations: Two-level games in the Anglo-Irish peace process. *International Studies
Quarterly*, 42, 545–565.

Tvinnereim, E. and Ivarsflaten, E. 2016. Fossil fuels, employment, and support for
climate policies. *Energy Policy*, 96, 364–371.

Ulama, D. 2015. *IBISWorld industry report: Electric power transmission in the
US*. Melbourne: IBISWorld.

UNEP. 2017. *The emissions gap report 2017*. Nairobi: UNEP.

UNFCCC. 2015. *Paris Agreement on climate change* [Online]. Available: http://
unfccc.int/files/essential_background/convention/application/pdf/english_paris_
agreement.pdf [accessed 12 August 2016].

UNFCCC. 2018. *Paris Agreement – status of ratification* [Online]. United
Nations. Available: https://unfccc.int/process/the-paris-agreement/status-of-
ratification [accessed 13 September 2018].

United States Congress. 2015. *H.R.1901 – PTC Elimination Act* [Online]. Available:
www.congress.gov/bill/114th-congress/house-bill/1901 [accessed 2 June 2016].

Unruh, G. C. 2000. Understanding carbon lock-in. *Energy Policy*, 28, 817–830.

US Chamber of Commerce. 2013. *State of American business, remarks by Thomas J.
Donohue President and CEO, US Chamber of Commerce* [Online]. Available:
www.uschamber.com/speech/state-american-business-remarks-thomas-j-donohue-
president-and-ceo-us-chamber-commerce: US Chamber of Commerce. Available:
www.uschamber.com/speech/state-american-business-remarks-thomas-j-donohue-
president-and-ceo-us-chamber-commerce [accessed 25 August 2016].

US Chamber of Commerce. 2014. Available: www.uschamber.com/speech/state-
american-business-2014-remarks-thomas-j-donohue-president-and-ceo-us-
chamber-commerce

US House of Representatives. 2015. *Testimony of Petr Gandalovič, Ambassador of
the Czech Republic to the United States, before the US House of Representatives
Energy & Commerce Committee, Subcommittee on Emergy & Power* [Online].
Available: file:///D:/Users/z3437188/Desktop/HHRG-114-IF03-Wstate-Gandalovic
P-20150709.pdf [accessed 25 August 2016].

US Senate Committee on Energy and Natural Resources. 2014. *Full committee
hearing: US crude oil exports: opportunities and challenges* [Online]. Available:
www.energy.senate.gov/public/index.cfm/hearings-and-business-meetings?
ID=4257c751-1911-4467-aaa5-0ff7863777fa [accessed 7 September 2015].

US Senate. n.d. *Party division* [Online]. Washington DC: United States Senate,
Historical Office. Available: www.senate.gov/history/partydiv.htm [accessed 2
February 2017].

USCAP. 2009. *A blueprint for legislative action*. Washington DC: United States
Climate Action Partnership.

Vernon, R. 1971. *Sovereignty at bay*. London: Longman.

Vogel, D. J. 1989. *Fluctuating fortunes*. New York: Basic Books.

Vogel, D. J. 1996. The study of business and politics. *California Management
Review*, 38, 146–165.

Volcovici, V. and Gebrekidan, S. 2014. Analysis: Lobbying fight over US oil exports
may be one-sided battle. *Reuters*.

Vormedal, I. 2011. From foe to friend? Business, the tipping point and US climate
politics. *Business and Politics*, 13, 1–29.

Vormedal, I. 2012. States and markets in global environmental governance: The role of tipping points in international regime formation. *European Journal of International Relations*, 18, 251–275.

Walke, John. 2012. Is your power company fighting in court against safeguards from mercury and toxic air pollution? Natural Resources Defense Council. Available: www.nrdc.org/experts/john-walke/your-power-company-fighting-courtagainst-safeguards-mercury-and-toxic-air [accessed 9 June 2016].

Waller, M. 2014. Wind power fights to regain subsidy; Lapsed tax credit is focus of argument. *San Angelo Standard-Times*, 11 January.

Wang, U. 2015. It's time to hurry up and beat a solar tax credit deadline. *Forbes*, 27 February 2015.

Wannier, G. E. 2010. EPA's impending greenhouse gas regulations: digging through the morass of litigation. 24 November 2010. *Climate Law Blog, Sabin Center for Climate Change Law* [Online]. Available: http://blogs.law.columbia.edu/climatechange/2010/11/24/white-paper-epas-impending-greenhouse-gas-regulations-digging-through-the-morass-of-litigation/ [Available 2 June 2016].

Ward, D. 2014. Republican Senator Chuck Grassley promotes wind power on US Senate floor. *Into the Wind* [Online]. 15 May. Available: www.aweablog.org/republican-senator-chuck-grassley-promotes-wind-power-on-u-s-senate-floor/ [Available 21 February 2017].

Warmann, J. 2015. Testimony of Jeffrey Warmann, CEO Munroe Energy Inc, on behalf of the CRUDE coalition to the US Senate Energy and Natural Resources Committee Crude Export Policy Hearing.

Weiss, L. (ed.) 2003. *States in the global economy: Bringing domestic institutions back in.* Cambridge: Cambridge University press.

Werner, T. 2015. A breakthrough year for solar. Available: https://us.sunpower.com/blog/2015/12/18/sunpower-ceo-tom-werner-weighs-vote-extending-solar-tax-credit/ [Available 14 April 2018].

Whieldon, E. 2014. *SNL FERC power report.*

Whitlock, R. 2013. *AWEA welcomes President Obama's commitment to wind power* [Online]. Available: www.renewableenergymagazine.com/wind/awea-welcomes-president-obamaa-s-commitment-to-20130214 [accessed 2 June 2014].

Witter, D. 2015a. *IBISWorld industry report 21111: Oil drilling and gas extraction in the US.* Melbourne: IBISWorld.

Witter, D. 2015b. *IBISWorld industry report 32411: Petroleum refining in the US.* Melbourne: IBISWorld.

Witter, D. 2015c. *IBISWorld industry report: Coal mining in the US.* Melbourne: IBISWorld.

Woll, C. 2008. *Firm interests: How governments shape business lobbying on global trade.* Ithaca: Cornell University Press.

Wood, J. and Shearing, C. 2007. *Imagining security.* London: Willan Publishing.

WRI. 2009. The USCAP blueprint for legislative action. World Resources Institute. Available: www.wri.org/blog/2009/01/uscap-blueprint-legislative-action [accessed 21 March 2013].

Yackee, J. W. and Yackee, S. W. 2006. A bias towards business? Assessing interest group influence on the US bureaucracy. *The Journal of Politics*, 68, 128–139.

Yergin, D. 2008. *The prize: The epic quest for oil, money and power.* New York: Simon & Schuster.

Young, K. 2015. Not by structure alone: Power, prominence, and agency in American finance. *Business and Politics*, 17(3), 443–472.

Zald, M. 1996. Culture, ideology, and cultural framing. *In:* McAdams, D., McCarthy, J., and Zald, M. (eds.) *Comparative perspectives on social movements: Political opportunities, mobilizing structures, and cultural framings.* Cambridge: Cambridge University Press.

Index

Page numbers in **bold** denote tables.

Printed and bound by CPI Group (UK) Ltd, Croydon, CR0 4YY

24/10/2024

01778279-0005